CROSSINGS

Cover: *Ely's Ferry*, by Lyme-school artist Carleton Wiggins (1848-1932). This scene shows a typical crossing — a wagon loaded with salt hay departing the Lyme shore. Essex and its congregational church can be seen in the background. Wiggins spent summers in Old Lyme from 1904 to 1917 when he bought a Victorian home near the Lieutenant River and moved here permanently.

CROSSINGS

Three Centuries from Ferry Boats to the New Baldwin Bridge

IAN HUBBARD
Introduction by Oliver Jensen

GREENWICH PUBLISHING GROUP, INC.
LYME, CONNECTICUT

ACKNOWLEDGEMENTS

I would like to thank the following ConnDOT and former ConnDOT individuals for their help in researching this manuscript: Robert Gubala, Dan Coffey, Alphonse Ciriello, Robert Gamache, William Sasser and also John Izzo - Rideworks, Inc. I would like to thank PBCS President Chris Reseigh for having confidence in me and giving me the opportunity to join PB and to work on this landmark project. I would like to thank all of those who worked on the Baldwin Bridge for the team spirit that has made this project a success and without which this book would not have been written - who wants to write about a bad project? I would like to thank Beverly Johnson for her assistance in putting it all together and to Mowry Mann, publisher, for believing that a book about bridges wasn't such a bad idea. Lastly, I want to thank my wife and family for their support through this and many previous bridges, which has uprooted the family more times than they care to remember.

— IAN HUBBARD

Grateful appreciation is given to JOSEPH GRIBBINS for his research and writing about the history of the lower Connecticut River area.

PICTURE CREDITS:

All photographs by the author unless otherwise noted.
All illustrations by Edith Twining.

FRONT COVER Rick Cooper
PAGE 6 Cottie and Jane Davison
PAGE 10 Florence Griswold Museum
PAGE 12 Connecticut Historical Society
PAGE 14 Old Saybrook Historical Society
PAGE 15 Lyme Town Hall
PAGE 17 Connecticut River Museum
PAGE 18 Oliver and Alison Jensen
PAGE 19 Connecticut River Museum
PAGE 20 Connecticut River Museum
PAGE 21 Connecticut River Museum
PAGE 22 Florence Griswold Museum
PAGE 23 (top) Barbara Maynard
PAGE 23 (bottom) Old Saybrook Historical Society
PAGE 24 Barbara Maynard
PAGE 25 (top) Barbara Maynard
PAGE 25 (bottom) George Goss
PAGE 26 George Goss
PAGE 28 (left) Rob Moore, Connecticut Department of Transportation (ConnDOT)
PAGE 28 (right) Barbara Maynard
PAGE 29 Middletown Press
PAGE 30 Rob Moore, ConnDOT
PAGE 32 Rob Moore, ConnDOT
PAGE 75 Johnny Gaines
PAGE 92 Johnny Gaines
BACK COVER (aerial) Rob Moore, ConnDOT

EDITING: Ceila Dame Robbins
ILLUSTRATIONS: Edith Twining
DESIGN: Tom Goddard Design
TYPOGRAPHY, SEPARATIONS & FILM ASSEMBLY:
Silver Eagle Graphics, Inc.

Library of Congress Catalog Card Number:
93-77717
ISBN: 0-944641-05-9

Produced and published by
Greenwich Publishing Group, Inc.,
Lyme, Connecticut

First edition: May 1993

CONTENTS

This is how the lower Connecticut River valley looked in August, 1973. The Earth Resources Experiment Project photograph was taken from a General Dynamics RB57-F aircraft flying at 60,000 feet. Clearly visible are today's roads, train tracks and geographical features, in addition to some from yesterday. The scars of earlier train tracks leading to the railroad ferry can be seen just above the present railroad bridge. Similarly, the old Route 1 approaches to the 1911 bridge can be located just north of the Baldwin Bridge. Route 1 leading to Old Saybrook can still be seen; so can the former tracks of the Valley Railroad running to its once-bustling freightyard and the steamboat dock at Saybrook Point.

Introduction

By Oliver Jensen

The New Baldwin Bridge is austere but magnificent — a wonder of wonders, a forerunner of the magical new century hovering just over the horizon. With its eight wide lanes, not to mention its walkway for the vigorous or the contemplative pedestrian, it has nevertheless gone up, it seems, almost overnight, ahead of schedule. Watching it rise from the vantage point of the old bridge, I have been particularly impressed by the swift progress of the superstructure; it must have reminded many an older man like me of the Gilbert Erector Sets of our youth, except a thousand times larger. Perhaps a million; these are big-number days.

We live in an age of marvels, multiplying every year, in medicine, power, communications, science, space; but the most visible to the human eye have to do with travel. Just two lifetimes the length of mine, which is 79, take us back to 1835, the age of the sailing ship, the stagecoach and the horse. The new craze was canals, especially the famous Erie, which carried my Norwegian great-grandfather, just off the boat, from Albany to Buffalo. On he went to the new little Ohio village of Cleveland, just settled by Connecticut men.

Connecticut briefly had a canal too, all the way from New Haven north to Northampton. Another was being surveyed from Norwich up to Worcester, where the other new marvel, a toy-like little railroad was approaching from Boston. Very sensibly, the Norwich & Worcester decided to become a railroad too, instead of a pokey waterway. By 1840 the trains were connecting with the new Norwich line of steamers, and my Yankee great-grandfather went to work for it. Not far away in Stonington, another little railroad had come in from Providence and Boston, built by George Washington Whistler, the noted engineer. (He was, by the way, the husband of the lady famous as "Whistler's Mother" in the painting by their artist son.) For a few years the citizens of busy Stonington, outraged by steam and smoke, forced the new railroad to haul its trains the last mile into town with horses, but you can't stop progress. The Stonington Line of "Mammoth Palace Steamboats" soon outshone the Norwich Line's flagship, the *City of Worcester*, called the "Greyhound of the Sound."

Another line of steamboats ran down the Connecticut River from Hartford as late as 1931, so that in the later days of the first highway bridge at the site of our New Baldwin Bridge, people in motor cars, stopped when the drawbridge opened, could witness the spectacular passage of these huge craft through a narrow passage. From Hartford, a competing railroad ran down to New Haven by 1839; you could take another steamboat line from there to New York. There was still another at Bridgeport, connected with the old Housatonic Railroad.

All these early solutions to the problem of moving people and freight between the two great cities of Boston and New York, in the days before good roads and automobiles, ran north and south down rivers or more easily-mastered valleys. To move east and west meant overcoming the many river mouths, bays, estuaries and marshes which the glaciers left on Connecticut's and Rhode Island's shores.

Nevertheless, by 1848 the first railroad reached New Haven from New York, and in that same year a hopeful New Haven & New London Railroad Company was chartered. With the muscles of thousands of new immigrant laborers, the line was completed to New London in 1851, a little over 50 very difficult miles, with curves which still plague Amtrak today. The Connecticut River was too much for them, however, and for 20 years the wooden railroad cars had to be shunted onto ferryboats to get across the river. There was a turning pit on each bank so that the locomotives could face in the right direction when they started back whence they had come.

In 1871, despite strong maritime and political opposition, the lower Connecticut River got its first bridge. The long, narrow structure carried just one track and had a huge draw. Meanwhile the Stonington Railroad had built a "Groton Extension" to a point just across the Thames River from the New London station. A giant ferry which could carry eight cars closed the gap until 1889 and the completion of what was then the largest double track steel drawbridge in the world.

All the parts of what would become the Northeast Corridor in New England were now in place, the hinterland linked to the cities, and everything expanding. Almost all the railroads were absorbed into the rich New York, New Haven & Hartford, then controlled by J.P. Morgan, who hated competition. To eliminate that untidy problem, he over time bought it all up, both steamboats and, later, trolley lines, and kept them all running on into the age of the automobile. He died in 1913. Consider the enormous changes in travel in all that time since 1835, the first of the "two lifetimes" I mentioned above.

The second of these arbitrary periods just happens to coincide with my birth in 1914; by the time I was five or six and could begin to notice the world around me, I assumed that progress was complete. My father took me to New York in the night boat from New London; wonderful little stateroom, breakfast in the saloon as Manhatttan and its giant Woolworth Building slipped by, home by night on the train from Grand Central and, for the last three miles, in a breezy open trolley car. We had no automobile yet, but an adventurous friend of father's took us on excursions in his open Franklin motor car. Automobiles crossed the Thames to Groton in a big ferry, the *Governor Trumbull.* Going in the other direction, down the mostly dirt or gravel Post Road to Lyme and Saybrook, I first remember the original 1911 Bridge across the Connecticut River, connecting two sleepy, pretty villages. The bridge seemed to go on and on, and the view was equally astonishing, up river and down for miles.

I was mystified by a trolley track, set into the decking on

the north side of the bridge, but not connected with anything on either end. It was an abandoned fragment of the Shore Line Electric Railway, which I learned in later years was once New England's longest interurban. At its full extent, it reached eastward from New Haven through all the shoreline towns, across this bridge, and on to New London, whence its branches, assembled from many earlier lines, extended to Stonington, Westerly, Norwich, and thence to Providence by one route and Worcester by another. There were some 250 miles in all. By further connections with other trolley systems you could ride all the way from New York to Boston — provided you had all the time in the world and $2.40 in small change.

Almost no one ever took that winding, back-country trip because the steam railroads were infinitely faster, although in some sections the big green wooden trolley cars would reach 60 miles per hour. At such high speeds the Shore Line Railway suffered two front-end collisions, one costing 19 lives and 35 injured. Combined with a strike and a failure of the power house, whose tall stack is still a landmark on the Saybrook side of the river, they put the line out of business, although some segments survived briefly under other managements.

The full system, a hopeless anachronism from the start but maintained as an expensive hobby by the financier Morton Plant, lasted just six years, from 1913 to 1919. Then the wires came down over the old bridge and the tracks were never used again. Most present residents of the Valley Shore area have never heard of these trolleys, and you can surprise them if you remark that the huge present project of electrifying the section of Amtrak between New York and Boston isn't really that new an idea — there was an electric railroad going by right here in 1913 (my lifetime plus one).

Crossing a great bridge over a famous, noble river is an exhilarating experience, a dramatic way of going from one place to a very different one — another city, or village, or nation. It is apt to stir the mind by the vistas that suddenly open up, and by one's own reflections on other crossings and other times. Progress, or at least change, never pauses for long, no matter how permanent the view around one may seem.

Bridges are monuments to man's skills and dreams, which this book, with all its interesting details, will make very clear. While progress never reaches a peak and stops, as I thought as a child, it may have to rest a while after completing the New Baldwin Bridge. Long may it stand!

— Oliver Jensen
April, 1993

Ellen Noyes Chadwick (1824-1900) painted this view of *Ferry Point, Old Lyme* from a spot near her home on Ferry Road. The view looks south towards the bustling wharves and warehouses of Ferry Point with Old Saybrook and its lighthouse in the background. Old Lyme had a busy maritime industry during the nineteenth century which no longer exists today.

CHAPTER 1 Crossings In Earlier Times

ANYONE who has ever traveled on the Connecticut stretch of I-95 between New Haven and New London knows about the Baldwin Bridge.

They may know it as a quick, scenic interlude in the sameness of the highway's forested flanks. They may remember the beautiful view the bridge presents of the Connecticut River — Saybrook Point and Old Lyme to the south, and the river stretching north to Essex and Lyme with hills folding down to the water and islands and boats for added interest. But what travelers will talk about, especially if they've driven through Connecticut on I-95 during summer weekends, is the unbelievable traffic jam that sometimes slows vehicles to a dead stop as far as Clinton and Madison to the west and East Lyme and Waterford to the east.

There are two main reasons for the weekend bottleneck at the bridge. First is the spectacular view, but equally disruptive is the additional volume of slow-moving local traffic from Routes 1, 9 and 156. Vehicles crossing the river from these roads must climb steep entry ramps with short acceleration lanes in order to merge with I-95 traffic. The present four-lane bridge just can't handle the volume.

This bridge that seemed a model of efficiency when it was built in 1948 became a major traveler's headache as early as the 1960s. Clearly, it was never intended to be part of a highway like I-95. Steadily increasing traffic on U.S. 1 during the postwar years led to the construction of the Connecticut Turnpike, which was completed in 1958. It was originally called the Greenwich-Killingly Expressway or the Blue Star Highway and was eventually absorbed into Interstate 95.

As the motorists of 1993 (including local residents who may cross the bridge several times a day) anxiously await completion of the spectacular eight-lane New Baldwin Bridge, this may be a good time to put this engineering marvel into historical perspective.

From sixteenth-century dugout canoes, driftwood rafts and colonial ferries to the Saybrook Bridge of 1911 and the Baldwin Bridge of 1948, crossing the lower Connecticut River has always posed challenges. Yet the people who have lived along its banks have always managed to cross the river either through their own ingenuity or, later, by hiring the very best to design and build ever-better ways to make the crossing.

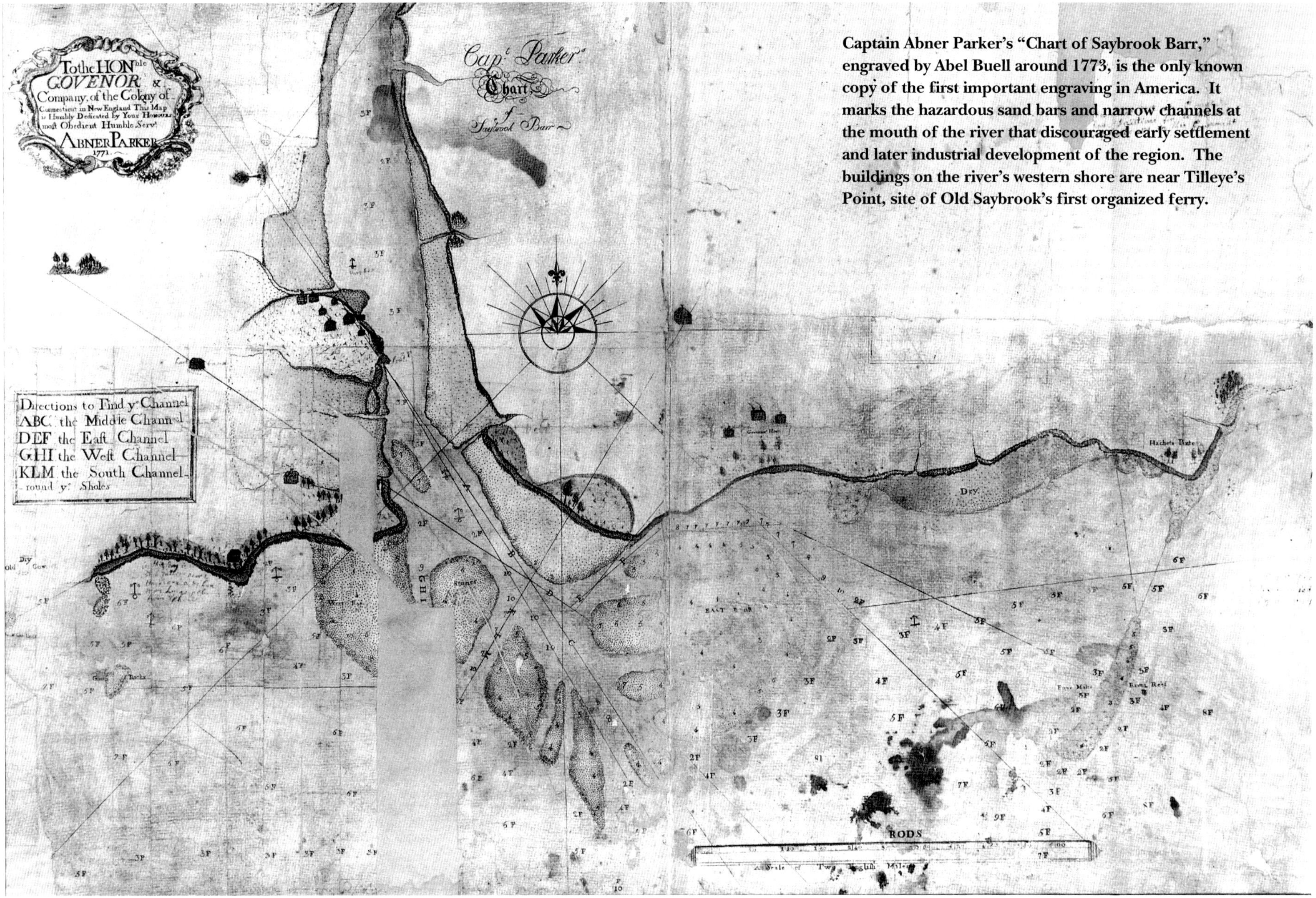

Captain Abner Parker's "Chart of Saybrook Barr," engraved by Abel Buell around 1773, is the only known copy of the first important engraving in America. It marks the hazardous sand bars and narrow channels at the mouth of the river that discouraged early settlement and later industrial development of the region. The buildings on the river's western shore are near Tilleye's Point, site of Old Saybrook's first organized ferry.

The Connecticut River bisects New England from just below the Canadian border all the way down to Long Island Sound, a distance of 410 miles. It provided the Native Americans of the region (and later European settlers) both a barrier and a bounty — a barrier to those wishing to cross from one bank to the other and a bounty of fish, navigable water and fertile bottomland.

Native Americans used simple rafts and dugout canoes to cross the Connecticut long before Dutch sea captain Adriaen Block explored the area in 1614. (Connecticut is the Anglicization of the Indian word "Quonehtacut," or perhaps "Quinatucquet" or "Quenticut," meaning long tidal river.)

The English sailed up the river and established Windsor in 1633. They settled Saybrook in 1635 and what is now Essex about 1645. In 1636, Hartford was settled overland from the Massachusetts Bay Colony by the Reverend Thomas Hooker's party, which crossed the Connecticut on driftwood rafts with their livestock and household goods.

During the 1650s, a flat-bottomed scow propelled by sail and oars could be hired to cross the lower Connecticut. Like all the river craft of the time, it was an informal arrangement rather than an organized ferry. Its route was reportedly between the mouth of Ragged Rock Creek in Saybrook and the mouth of the Lieutenant River in what is now Old Lyme. (Lyme was considered part of Saybrook Colony until 1665, and the separation of Lyme into the towns of Lyme and Old Lyme didn't happen until 1885.)

In 1662, the Connecticut General Court gave settlers at Saybrook "liberty to set up a Ferrey at Tilleyes Point, and to take 12d. for a man and horse, and 6d. for a single person." Tilleyes Point (sometimes spelled with one or no "e") was just above the present railroad bridge in Old Saybrook along what is still called Ferry Road. The town of Saybrook later built another landing at Ferry Point for steam ferries, further up the river above the present highway bridge. Other ferries used Saybrook Point as their western terminus.

On the opposite shore, the ferry landings were at the end of a pathway which eventually became Ferry Road in Old Lyme and just to the north at a tiny peninsula known as Ferry Point. (More often than not, the ferries from Saybrook landed in Lyme wherever the wind and currents took them.) Passing through Lyme, travelers had the further inconvenience of crossing the Lieutenant River until the town's governing Ecclesiastical Society approved the construction of the first "Great Bridge" across the Lieutenant in 1732.

The crossing between Saybrook and Lyme was the first of the river's authorized ferries, although previous ferry services appear in the records at Windsor in 1641 and at Rocky Hill in 1655. The Rocky Hill-Glastonbury ferry is still in operation and, according to the best available records, is the oldest continuously operated ferry service in the U.S.

Ferries were a public service and a commercial necessity, taken seriously by the authorities of the colony. In the last half of the 1600s, the Connecticut River Valley developed a rich trade in agricultural products — pork, beef, wheat, rye, barley, corn, wool, apples — and this cornucopia was shipped south to New York and north to Boston and increasingly to the West Indies.

The shipment of these commodities increased the need for reliable ferries. Whether barrels of apples traveled from Lyme to Saybrook or from Saybrook to Jamaica, they made

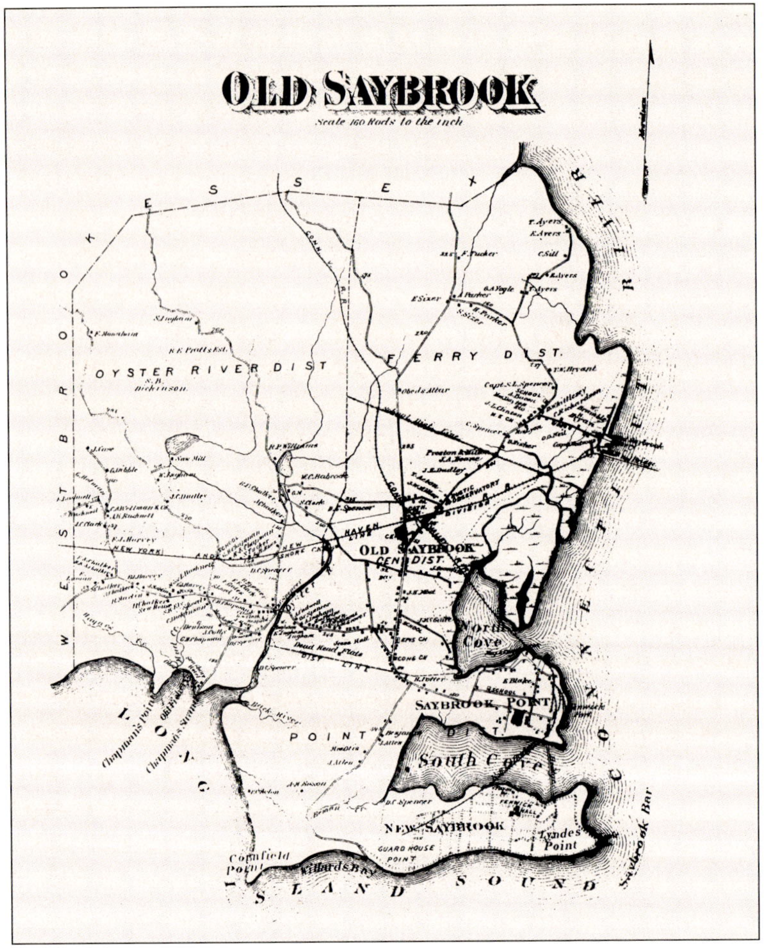

This 1874 map of Old Saybrook was drawn much later than the days of the Whittleseys and Dudleys, but their ferry landings were still in use. Their wharf was at the end of Ferry Road in the aptly named Ferry District, just north of the railroad bridge built in 1870. Another ferry left from Saybrook Point which, by the time this map was drawn, had a busy rail yard for freight moving between steamboats and the Connecticut Valley Railroad.

most of their journey by water. It is sometimes difficult for us, in our own world of superhighways, turnpikes and eight-lane bridges, to remember that in a year like 1704 there were hardly any roads; what existed then were trails wide enough for a horse.

The life of the ferrymen was difficult, at best. The ferry business was uneven in terms of traffic, uncomfortable and even hazardous on stormy days and in the winter. Ferrymen had to maintain their boats and landings and be on duty from sunup to sundown. It helped to have an inn as well as a ferry.

In 1737, Jabez Chapman, descendant of Captain John Chapman who had started an East Haddam ferry in 1664, petitioned the General Court to let him build an inn. He complained that his ferry was not competitive with other crossings because the nearest hospitality was three miles from his western landing and more than three miles on the east. Jabez Chapman got his wish, building a tavern and lodging he called the Riverside Inn on the site of the present Gelston House in East Haddam.

In the eighteenth century, as traffic increased on the Indian trails and as commerce increased in the ports of the lower river, additional ferries were started by local entrepreneurs and authorized by the General Court. Brockway's ferry began between Essex and North Lyme in 1724. Elsewhere in Lyme, Ely's ferry and Selden's ferry seem to have unofficial histories dating back to 1760. Both were upstarts and rivals of the Saybrook-Lyme crossing in the increasingly competitive (if not exactly lucrative) business of ferrying people and goods on the lower river during the 1700s.

Ferry service was established between Haddam and East Haddam in 1741 and 1749 — both new ferries in competition with the Chapman ferry and inn for traffic west from Norwich and down from Hartford. The first ferry between Chester and Hadlyme, begun in 1769, still plies the river today, now carrying cars and trucks instead of farm wagons and livestock on the hoof.

Keeping a close eye on the ferrymen

Meanwhile, between Saybrook and Lyme, there was occasional trouble with the ferry. In 1706, while the Whittleseys and Dudleys still operated the original Tilleyes Point ferry, another ferryman was briefly authorized by the General Court: "Upon the petition of Timothie Mather of the town of Lyme, this court grants him libertie to keep a boat and to transport travailers as there shall be occasion over the River of Connecticut near the mouth of it, he taking no other fees than what is already stated by lawe for that ferry, and attending the duty required of ferrymen."

Apparently the travelers really were travailers — that is, they were unhappy with Mather's service — because Mather's ferryman privileges were revoked shortly by the Court "for weighty reasons laid before them." The weighty reasons may have included overcharging — a frequent gambit for the ferrymen and a frequent problem for passengers and politicians. Ferry fares were set by law for each ferry, and were adjusted from time to time.

In 1719, the General Court was made aware of the need for adequate wharves and approach roads for the ferry landings at Lyme and Saybrook, and they appointed John

The Lyme side of the river can be seen in this 1850 map. The eastern landing of the Tilleyes Point ferry was at Ferry Point, located at the end of Ferry Road. Ferry Road was the first road built by the town of Lyme, one year after it split from Saybrook Colony in 1665. Further north, Ely's Ferry provided crossings where the river narrows between Essex and Lyme. Carleton Wiggins's painting, *Ely's Ferry*, is on the cover of this book.

A 1704 TRAVELER'S DIARY

HERE IS what Boston schoolteacher Sarah Kemble Knight experienced at the Connecticut River in October of 1704 on a ride from Boston to New York: "Wee advanced on towards Seabrook. The rodes all along this way are very bad, Incumbred with Rocks and mountainos passages, which were very disagreeable to my tired carcass...From hence we went pretty briskly forward, and arriv'd at Saybrook ferry about two of the Clock afternoon; and at an Inn to Bait (foreseeing we should not have such another Opportunity till we come to Killingworth.)

"Landlady come in, with her hair about her ears, and hands at full pay scratching. Shee told us shee had some mutton which she would broil, which I was glad to hear; but I supose forgot to wash her scratchers; in a little time she brot it in; but it being pickled, and my guide saide it smelt strong of head sause, we left it, and paid sixpence a piece for our Dinners, which was only smell."

"Bait" is an old word for taking refreshment. "Head sause" is, well, unpalatable. The scratching landlady would have been one of the Whittleseys — old Ruth Whittlesey or one of her daughters. John Whittlesey married Ruth Dudley, and the ferry between Saybrook and Lyme was operated by John Whittlesey and his brother-in-law William Dudley. Both were given land by the town fathers of Saybrook in exchange for their ferry service to the community, and both families continued the business into the new century. An inn was part of the enterprise — and probably the paying part.

Hamlin and Richard Christophers to study the problem. The Whittleseys and Dudleys owned the ferry rights equally. Mary Dudley and her son William had begun to build a wharf on the Saybrook side of the river, and Stephen Whittlesey had begun to build a causeway and road from the Dudley's wharf to his house. It was therefore decreed that Whittlesey would reimburse the Dudleys one half the cost of the work on the wharf and that the Dudleys would reimburse Whittlesey one half the cost of his road improvements. The Dudleys and Whittleseys would share the ferry, taking over its operation in alternate years.

This Solomonic decision was apparently acceptable, and the families remained partners. In 1732, the General Court was petitioned by William Dudley and Ambrose Whittlesey, the Saybrook ferrymen, for a fare increase in recognition of the cost they had borne in maintaining their wharves. This was fair, for damage by ice could be considerable, even for stone wharves and causeways over marshes. The partners also complained that officers of the government were allowed to ride the ferry free of charge, and the fare increase was partly meant to satisfy that grievance. Perhaps this was protest against the Tory government and an early hint of the insurrection to come. The fare increase was granted — 13d. for man, horse and load in December, January and February; and 9d. during the other months.

It's interesting to note that the principal fare in 1732 was for man, horse and load rather than for a man and horse as it was in 1662. Some of the original trails were now wide enough for wagons, and commerce was on the move. Although the ferrymen occasionally won price increases, they could also be fined for not keeping up their boats and landings. In 1752, the General Court was made aware of "great complaints made that the ferry over Connecticut River at Saybrook has for a considerable time been very much neglected, and consequently the passage over said river is often attended with great difficulties and dangers."

Growth of the ferry business

One more problem with the Saybrook-Lyme ferry came up in 1760, after two boats began to be required. Passengers complained of "the negligence of the ferrymen in suffering both the boats frequently to remain long on the same side of the river, and thereby occasioning great delay to travelers." The result was a new law — "That the ferrymen or tenders of the ferries at New London or Saybrook shall carefully and diligently attend the convenience of passengers; and to that end they shall not at either of said ferries suffer said two boats to lie at the same time on the same side of the river...."

There was also a new schedule of fares for both ferries, including some reductions for old categories such as "man, horse and load," and a few new categories such as "wheelcarriages." Many roads were now good enough for carriages, in which prosperous farmers rode instead of wagons.

The Saybrook ferry crossing was on what was called the Lower Road from Boston to New York by way of Rhode Island — one of three Boston Post Roads that merged at New Haven. The others were the Upper Road, which came through Worcester, Springfield and Hartford; and the Middle Road, which came through northeastern Connecticut to Hartford and then to Wethersfield, Berlin, Meriden, Wallingford and North Haven.

Ferry fares were not the only things changing in the Connecticut River Valley in the last half of the eighteenth century. The farms and towns continued to prosper; the traffic in goods and passengers increased on the river and up and down the coast from Newfoundland to the West Indies. As commerce increased, so did the size and sophistication of ships built along the banks of the lower Connecticut — some of them the celebrated packets and clippers that would travel the world over.

Soon there were bridges. The first bridge over the Connecticut River crossed at Bellows Falls, Vermont, in 1785. It was a covered bridge 543 feet long and 50 feet above the water. Another covered bridge crossed at Windsor in 1796.

The first toll road completed in America was the Mohegan Road between New London and Norwich, an enlarged Indian trail opened in 1792. This road was state-funded, but many others were built by private interests

An early sail-powered ferry crosses the lower Connecticut, looking in this view like an Egyptian dhow. While there were several ferry landings on each side of the river, boats often ended their crossings wherever winds, tides and currents took them.

chartered by the state and often funded by lottery. Between 1795 and 1853, 121 of these turnpike or toll-road franchises were granted in Connecticut. The greatest surge of inter-town road building came in the years after 1820. The Connecticut crossings had never been busier.

Steamboats come to Connecticut

The *Fulton* made history in 1815 when she became the first steamboat to make the journey up the long tidal river to Hartford. Steamers soon began running three times a week between Hartford and New York, and travel on the river was never the same again.

Meanwhile, a few improvements came to the doomed technology of the ferries. Most New England ferries were propelled by sweeps until steam came along, although some were pulled across smaller waterways by ropes, or rigged to be driven diagonally downstream along a cable by the push of the current and upstream by the incoming tide. The rope ferry across the Niantic River left its name on Rope Ferry Road in Waterford. And some ferries were powered by sail when wind conditions permitted. An 1819 gazetteer reports of Lyme that "there is a ferry, accommodated with sail boats, upon the Connecticut River, maintaining a communication between this town and Saybrook...."

The *Lady Fenwick*, crossing from Saybrook Point to Lyme in the last half of the nineteenth century, was one of the lower Connecticut's first steam-powered ferries. The historic Ferry Tavern is just above her smokestack. Today, another *Lady Fenwick* offers sightseeing excursions on the lower Connecticut during the summer months.

In the last half of the nineteenth century, steam ferries finally arrived at the Saybrook-to-Lyme crossing. The *Lady Fenwick* provided faithful service until 1904 when she was replaced by the *Colonial* which was struck by lightning and burned during the summer of 1911. Because the *Colonial* was carrying 14,000 vehicles per year, two substitute ferries were quickly provided to handle summer traffic. But their services were short-lived; on August 24, 1911, the first bridge for motor vehicles was completed between Old Lyme and Old Saybrook.

The dawn of the railroad era

While the new highway bridge was a first for automobiles in the area, trains had been crossing the lower Connecticut by bridge since 1871 and by ferry for two decades before that. The story of railroading in Connecticut is interesting, particularly because it reminds us that politics has always played an important role in the building of bridges.

Private companies and local civic groups began obtaining

TRIALS OF THE FIRST RAILROAD BRIDGE

WHEN THE state legislature finally approved the bridge across the lower Connecticut in 1866, the New Haven and New London Railroad hired Phineas Earl Peck of Woodbridge to supervise the construction. He and his crew of local craftsmen built the Howe truss bridge by hand. It was made in six sections with a center section that swung aside for river traffic. Observers noted proudly that all the wood came from Connecticut forests.

Opposition to the new bridge had been intense and continued long after construction started. In a *New Haven Register* article of 1947, Peck's son remembered his father telling him of the big barges coming down from the Portland stone quarries and intentionally ramming the bridge. "These huge scows always became unmanageable in the vicinity of the bridge site. And many a bridge pier. . . was torn away from its hard-won mooring in the stream when a stone barge buffeted it. Dad was a scrapper, a big man and a good builder. In the end, he got it up."

Evidence of Peck's fortitude came again on the bridge's opening day. Just as the first train was about to go over the bridge, it stopped on the Saybrook side in front of a crowd of spectators and railroad executives. The engineer and his crew jumped off, fearing that crossing the bridge was too risky. Unwilling to see his bridge humiliated, Peck climbed into the cab and called for volunteers to join him.

One of his fellow bridge builders swung aboard and was followed by one of the railroad's top officials, Mr. Scranton. "By gad, Peck," he said, "if you're going on this thing, I'm going, too."

As the *Register* reported, "the crotchety little engine trundled out onto the bridge, rolled along its wooden ties, passed the first section, bumped in mid-stream onto the swinging platform and continued on in triumph to the Lyme shore."

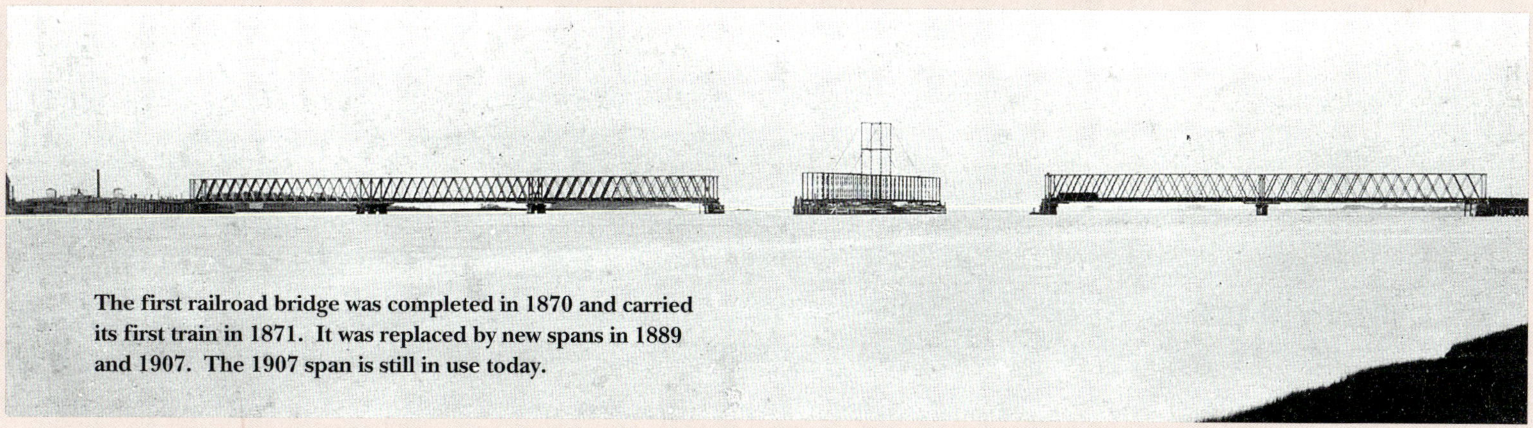

The first railroad bridge was completed in 1870 and carried its first train in 1871. It was replaced by new spans in 1889 and 1907. The 1907 span is still in use today.

Before the 1870 railroad bridge was built, the ferry *Shaumpishus* crossed from Saybrook to Lyme carrying railroad cars and their passengers. From 1852 until 1871, railroad cars were pushed onto ferries for the crossing. On the other shore, a new locomotive pulled them on their way. The cabins to the port and starboard of the *Shaumpishus* provided food and rest for passengers.

charters to build railroads in the 1830s. Limited resources, however, demanded that their routes avoid major obstacles like rivers and rough terrain. Yet to compete with steamboats for both freight and passengers, they needed long-distance routes; important among them was the shoreline route from New York to Boston.

In 1848, the New Haven & New London Railroad was chartered to build a 50-mile line from Fair Haven to New London The new route would eventually connect with railroads already established from New York to New Haven and from Stonington to Boston.

The Fair Haven to New London route opened in 1851 and by 1858 the connection from New London to Stonington was complete. But a trip along the coast of Connecticut required two unusual interruptions — railroad cars were shuttled onto ferry boats to cross both the Connecticut and Thames rivers. On the opposite bank, a new engine would assemble the cars and the journey would continue.

Charles Dickens traveled this route in 1868 and later described the experience: "...two rivers had to be crossed, and each time the whole train is banged aboard a steamer. The steamer rises and falls with the river which the railroad don't do, and the train is banged uphill or banged downhill. In coming off the steamer at one of these crossings yesterday, we were banged up to such a height that the rope broke and the carriage rushed back with a run downhill into the boat again. I whisked out in a moment, and two or three others after me, but nobody else seemed to care about it."

The inconvenience Dickens and other travelers suffered was a result of the powerful steamboat lobbyists who managed to prevent railroads from building bridges across rivers.

The steam ferry *Colonial* started crossing the Connecticut in 1904. The 1907 railroad bridge can be seen to the right, along with Lyme's busy waterfront including the warehouses on Ferry Point, the subject of Chadwick's painting on page 10. The *Colonial's* route went from Ferry Road in Old Saybrook to the Ferry Tavern on Ferry Road in Old Lyme.

In the early days of aviation, an intrepid 1921 barnstormer took this aerial shot of the lower Connecticut showing the Saybrook Bridge in the background with Ferry Point and the railroad bridge below. The old railroad ferry landing is still visible jutting out just above the railroad bridge. Ten years earlier, the Saybrook Bridge became the first highway connector between Old Lyme and Old Saybrook. The open land was typical of both sides of the river during earlier times. Land was cleared not only for farming; trees were sold for shipbuilding and commerce.

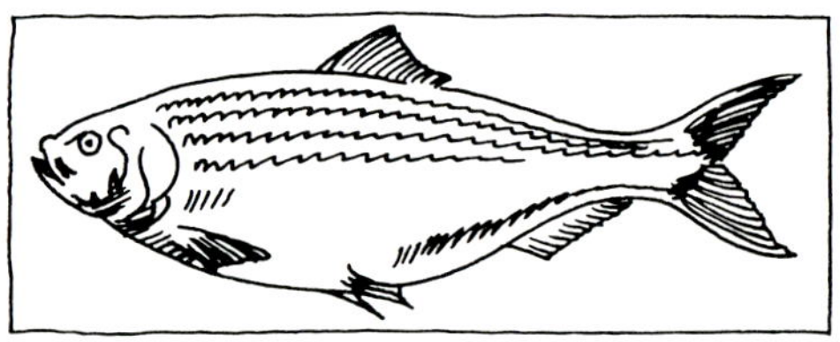

SHAD

They successfully defeated an attempt to build the Saybrook bridge in 1866 and halted another railroad bridge between Middletown and Portland in 1867. But time and technology were against the steamboat interests. The first bridge across the river between Saybrook and Lyme was a single-track railroad bridge that carried its first train in 1871. That bridge was replaced by new spans in 1889 and again in 1907, when the present double-track drawbridge was built.

Once the first bridge was in place, the steamboats quickly lost passengers and freight to the faster trains. The railroad interests in their turn and with the same well-justified fears, opposed the highway bridges in the new century. Also opposed to bridges were the shad fishing interests. They feared additional obstructions in the river would deter the yearly migration which so enriched the local economy.

Crossing the Connecticut by automobile

After centuries of crossing by canoe, raft and ferryboat, the citizens of Saybrook and Lyme were finally joined together in 1911 by the erection of the Saybrook Bridge. Though the horse and buggy days weren't quite over, the quickly spreading phenomenon of the automobile created the need for the bridge. In 1911, there were 21,371 cars registered in Connecticut. The year before the bridge opened, 14,000 cars and trucks had crossed the river aboard the old *Colonial.*

Construction was completed about 17 months after the state assembly passed the act permitting its start. It was expected to cost $500,000, but came in about $25,000 under budget. And it was controversial — a toll bridge built despite a state policy that favored free bridges. The toll was five cents, and it remained a nickel until 1923 when the bridge became free.

The Saybrook and Lyme Connecticut River Bridge Commission, comprised of three distinguished citizens of the region: D.A. Blakeslee of New Haven; A.N. Shepard of Portland; and James H. Day of Old Saybrook, hired three different firms to design and construct the bridge. The Holbrook, Cabot & Rollin Corporation of Boston built the substructure. Boller & Hodge of New York City designed the superstructure, which was built and erected by the Pennsylvania Steel Company of Steeltown, Pennsylvania.

Though not a standard-bearer in terms of new construction methods, the Saybrook Bridge was made with the sturdy ingenuity of the time. It opened to traffic on August 24, 1911, with what the *New Era* of Deep River called "a monster automobile parade" of 500 cars, plus many more automobiles on the sidelines, all decorated with flags, ribbons, bunting and flowers. Former Governor and Senator Morgan G. Bulkeley rode in a car covered with purple wisteria; another car was outfitted like a ship with nautical flags strung from her spars.

The bridge-day parade began at Old Saybrook's town hall. It followed the old Ferry Road to the new bridge, crossed the river, wound through Old Lyme and back to the bridge for speeches and ceremonies. Accompanying the cars were wagons, fancy carriages, schoolchildren and their teachers, and citizens from near and far. The *Penny Press* account of the day included the information that, "Word of the program apparently traveled to many near by metropolitan centers as it attracted the largest number of peddlers and fakirs ever recalled in the memory of the oldest inhabitant."

When the first highway bridge opened on August 24, 1911, the towns of Old Saybrook and Old Lyme hosted an immense automobile parade to celebrate. The first car shown above on Main Street in Old Saybrook is trimmed with American flags. Governor Bulkeley's wisteria-covered car is third. Below, Dan Spencer and family with Mrs. Durham ride in their imaginative dirigible car. The elaborate three-day celebration reflected awareness of the importance of the new bridge.

THE BUILDING OF THE 1911 BRIDGE

Old Saybrook's and Old Lyme's new connection was a plain but impressive job of steelwork that stretched 1,800 feet between the river banks, where it was met by more than a mile of macadamized road on each side. Over the channel, a two-arm Scherzer drawbridge opened electrically and gave a 200-foot-wide clearance to vessels — which in 1911 included three-masted schooners up from Newport News with cargos of coal.

To decide where the draw channel should be located, two rowboats carrying red and white flags at day and white lights at night were anchored at the edge of the proposed channel. All tugboat and steamboat captains using the river were then asked to observe their location and make suggestions to the engineer for the best possible lanes. The rowboats were shifted from time to time until the shipping interests were satisfied.

The bridge was built on 12 piers placed in from 15 to 40 feet of water on piles supporting timber grillages on which the masonry was laid. The abutments, 39 feet high, also rested on piles, over and around the tops of which were placed a concrete footing course. The piles, mostly chestnut and hemlock, were from 25 to 40 feet long below the cut-off elevation and up to 77 feet below high tide.

The grillages were made up of five levels of timbers with alternate directions in each layer, measuring approximately 22 feet by 52 feet. The piers were constructed above them to a point a little below mean low water where the ashlar started. They were made of Portland brownstone rubble laid in large blocks and well bonded. The ashlar consisted of Portland brownstone backed with concrete and capped with granite coping.

The superstructure included about 2,200 tons of steel. The trusses were of the deck Warren type with a bottom chord panel length of nearly 40 feet and a sub-panel length at the top chord of about 20 feet. The floor joists were 20-inch I-beams, spaced about 10 feet apart, which rested directly on the top chords.

The floor surface of the bridge was two-inch spruce plank laid transversely on top of three-inch creosoted yellow pine laid diagonally. The roadway was 24 feet wide between the wheel guards, and a trolley track was laid near one side of the roadway. The bridge was designed for a 60-ton trolley car.

The trolley connections in 1911 were south from Old Saybrook to Stony Creek on the Shore Line Electric Railway Company System, and north from Old Lyme to New London aboard the Connecticut Company's trolleys. From Stony Creek, trolley connections went all the way to New York: from Groton, a shoreline trolley ran to Westerly.

Electricity provided by a 37 horsepower motor on each arm was used to operate the draw, which could be fully opened in 70 seconds. Three men were employed to operate it, each man taking an eight-hour shift which also included collecting the tolls. The head bridge tender was on call virtually 24 hours a day, thanks to the eight-room cottage built for him rent-free near the west abutment.

In the first years of operation, horses and buggies were as likely to cross the bridge as automobiles.

A pundit in the New London *Day* wrote: "The opening of this bridge means far more than the casual thinker realizes. It means a diversion toward New London of highway travel in large amounts; it means that into the city from the west is now open an unbroken highway, over which traffic may come for the upbuilding of this city." And indeed for all of eastern Connecticut. And so the 1911 bridge served the motoring public and the people of the state well for 36 years.

By 1947, however, it was carrying a daily average of 6,500 cars and trucks. It had reached its limitations in both volume of road traffic and number of bridge openings — mostly for sailing yachts in the 1930s and 1940s. As the *New Era* noted in 1948: "Besides being two lanes too narrow for the traffic it bore, it had become a bottleneck when the lift span was raised for river traffic. Particularly was this the case during the summer weekends."

The raising of the bridge for river traffic did not sit well with the motoring public of the 1940s, especially on a hot July Sunday with bumper-to-bumper traffic on U.S. 1 from Rogers Lake to the bridge on the east side and through Old Saybrook nearly to Indiantown on the west.

Lieutenant Carroll Shaw, commander of the Westbrook Barracks of the Connecticut State Police, remembered the angry remarks of motorists in 1948: "Officer, this is a disgrace!" "Special privileges for the boat owners!" "Discrimination against the car owner!" "I'll write to the Commissioner!" "I'll see the Governor!"

A proposal for the first Baldwin Bridge had actually been put forth a decade earlier, in the late 1930s. Among other problems, the draw of the old bridge was constantly getting stuck and the wood timbers of the roadbed needed replac-

A sailboat passes underneath the Scherzer arms of the Saybrook Bridge near the eastern shore of the river. A 37-horsepower motor activated each arm, which could be fully opened in 70 seconds. Truman W. Brown, senior operator of the bridge, worked the draw for its final eight years. He reported to the *New Era* that between July 1, 1947 and June 30, 1948, the draw was opened 4,552 times, including 30 times during one eight-hour period on a summer Sunday.

ing. But the project was delayed by World War II, and Connecticut had to make do with repairs on the old bridge so that it could safely support vital munitions and military traffic on Route 1.

By 1945, postwar building projects were being presented to the state's General Assembly. The state representative from Old Saybrook, George A. Maynard, and Lyme's representative, E. Lea Marsh, Jr., sponsored the bridge bill and later became part of the five-man Old Lyme-Old Saybrook Bridge Commission which handled preliminary cost surveys and administered the actual construction of the bridge.

At local meetings, there was quite a bit of often heated discussion about where the bridge and its approaches should be located, what kind of bridge should be built and whether the allotted funds would be enough to cover all costs, especially since the price of labor and materials in postwar America were constantly rising. There was some talk of a suspension bridge; and one resident, a former war engineer who had built crossings on the Rhine, Colonel Needles, suggested a tunnel under the river, but this was found to be three times as expensive.

Finally, with the help of a $20,000 contribution from the town of Old Lyme, construction officially began on March 31, 1947. The design chosen by the bridge commission was a high-level, steel girder structure, consisting of 12 spans — three of 144 feet, two of 168 feet and seven of 240 feet, making a total length of 2,448 feet.

The new bridge of 1948 solved traffic problems on U.S. Route 1 and on the river, providing four lanes for vehicles and a clearance above the river of 85 feet at average low water, summer stage. Like the 1911 bridge, it had a horizontal clearance of 200 feet across the channel, but unlike the 1911 crossing there was no drawbridge to stop traffic. The 85-foot clearance was a compromise — some shipping interests wanted much more. (Indeed, Governor Chester A. Bowles had a sailing yacht which didn't quite clear the bridge at high tide. Rather than use his considerable influence to raise the clearance, he simply had his mast cut down a few feet.)

The new bridge opened on December 4, 1948, and the ribbon was cut by U.S. Senator Raymond E. Baldwin, the former Connecticut governor whose administration had begun the project and whose name was given to the bridge.

There was a parade in 1948 even bigger than the parade of 1911 — fire engines, school bands, the U.S. Coast Guard Academy Band, soldiers from Fort Wright on Fishers Island, marines from the submarine base, the 102nd Army Band, a platoon of Connecticut State Police (for once enjoying bridge duty), the Connecticut National Guard, Boy Scouts, Girl Scouts, American Legion marchers, fife and drum corps from Deep River, Essex, Moodus and Chester, and hundreds

By 1947, the two-lane bridge had become a traffic bottleneck and construction began on the four-lane Baldwin Bridge. Cofferdams and piers can be seen in this snapshot.

THE BUILDING OF THE 1948 BRIDGE

THE DESIGN of the structure was under the supervision of W.G. Grove, engineer of structures for the Connecticut Highway Department, and John Willis, engineer of bridge design, with Philip Keene, advisor and soil mechanics engineer. The Bridge Commission engaged Howard, Needles, Tammen and Bergendoff, consulting engineers of New York City, as their consultants.

The project, including approaches and four local-road overpasses, was divided into 10 contracts, totalling $5,528,692, including the cost of land acquisition. The cost of the substructure for the main bridge was $1,380,739, with Merritt-Chapman and Scott Corporation of New York and New London as the contractor.

Actual construction commenced on March 31, 1947, and was opened to traffic on December 4, 1948, only three days later than the scheduled time of December 1, 1948.

American Indians, who had been a major part of the work force on such projects as the Empire State Building and the Triborough Bridge in the 1930s, were hired as ironworkers. Among them was Edward Kezer, 43, of Campton, New Hampshire, who was the bridge's only fatality. Kezer and another worker lost their balance while jockeying steel plates into position on a 40-foot-high pier. Though his co-worker landed safely in the water, Kezer sadly hit one of the construction barges and died at Lawrence and Memorial Hospital two hours later. A bronze plaque placed on the bridge in his memory will eventually be transferred to the New Baldwin Bridge.

As in the first bridge, timber piles were considered and it was thought that they might be superior to steel H-piles in terms of their lower cost per foot; however, they could carry considerably less load per pile than the H-piles and hence larger footings, with more excavation and concrete, would be required. At some piers, the contractors were allowed to bid both types of foundations. In most cases, the alternate bids received were slightly lower for the steel piles than for the timber piles.

The concrete was handled by the contractor on a self-contained floating plant and a 95-foot elevator frame. The concrete was placed into the forms by means of swiveled chutes from the towers. This method proved to be satisfactory and was used until a fire destroyed the plant, after which an improvised plant was hastily put together to complete the balance of the concrete operations.

On top of the tremie, a reinforced concrete foundation beam was placed to receive the masonry and concrete pier shafts. The masonry consisted of a heavy granite stone masonry veneer caulked with lead wool. The pier shafts consisted of two concrete columns, capped by a concrete cross beam supporting two rocker bearings.

The superstructure was of a twin steel plate girder construction with floor beams at 24-foot centers and stringer beams supporting the reinforced concrete deck.

This type of construction was very typical of the era but is seldom used today. The preferred steel option today would involve multiple support beams in place of the two plate girders so that if a problem occurred in one of the support beams, the load could be carried through cross members to the remainder of the structure.

Construction involved approximately 14,000 cubic yards of excavation, 22,500 cubic yards of concrete, 950,000 pounds of steel rebar, 4,561,000 pounds of structural silicon steel, 7,194,000 pounds of structural carbon steel and about 90,000 field rivets.

Typically heavy traffic moves across the old bridge as the Baldwin Bridge nears completion. The 100-ton traveling crane used to erect most of the superstructure can be seen in the background. The toll gate on the Saybrook side collected a 10-cent toll for 20 years. Shortly after the Baldwin Bridge was completed, the 1911 bridge was demolished. Approaches to the 1911 bridge can be clearly seen here as well as in the high-altitude view on page 6.

of schoolchildren — all on the march from Old Lyme to the toll plaza on the Old Saybrook side where the ceremonies proceeded briskly on a brisk day.

It was, indeed, a great day for the two communities; for the politicians and engineers who had planned and completed the bridge; and for the tourists, truckers, commuters and commercial travelers who would use the bridge for the next 44 years. The newspaper editorial writers proclaimed that the bridge would manage all traffic over the Connecticut River for the foreseeable future.

But none among the celebrating citizenry could have foreseen the need for today's "New" Baldwin Bridge, the latest in the evolution of crossings of the great river once called Quonehtacut; a natural progression that began with simple dugout canoes.

RAYMOND EARL BALDWIN

THE BALDWIN BRIDGE was named for a man who made extraordinary contributions to Connecticut. Raymond Earl Baldwin is the only person in the state's history to serve as governor, United States senator and chief justice of the state supreme court.

It is a measure of the high esteem in which he was held that the bridge was named for him not after his death, but while he was still an active and vigorous public servant of 55.

Though Baldwin was born in Rye, New York, in 1893, he spent virtually his entire life in Connecticut. He grew up in Middletown and graduated from Wesleyan University, attended Yale Law School (interrupted by a three-year stint on a destroyer during World War I) and settled in Stratford, practicing law in New Haven and serving as the Stratford town attorney and town court judge.

His political career began in 1930 with his election to the state's house of representatives. After only three years, he became house majority leader. A founder of the liberal branch of Connecticut's Republican Party, Baldwin moved quickly through the hierarchy of the state government. In 1938, he narrowly won an unexpected victory over the incumbent Democratic governor, Wilbur L. Cross.

Baldwin was a fiscal conservative who espoused the principle, "There is no substitute for a good job in private industry," at a time when Roosevelt's New Deal dominated American business and politics. But he was also socially progressive, heading a group of governors who encouraged the formation of the United Nations after World War II and starting the Inter-Racial Commission, the first state civil rights organization in the country. Wendell Wilkie even wanted Baldwin as his running mate in 1940 when he ran for president against F.D.R., but finally chose Charles McNary of Oregon for geographic reasons (Wilkie, of course, lost in a Roosevelt landslide).

In 1946, Baldwin was on the verge of retiring from politics when friends persuaded him to run for the United States Senate. He won handily, but his experiences in Washington were not entirely happy. Frustrated by the GOP's apparent disinterest in attracting young and independent voters, particularly with regard to lowering high prices and dealing with the housing shortage, he shocked Washington and Connecticut by resigning from the Senate in 1949.

Many at the time felt he was abandoning his own party, but in fact his main reason for leaving was to pursue a lifelong dream. The former Yale law student and Stratford town attorney was appointed by Governor Chester Bowles to sit as an associate justice on the state supreme court. In 1959 he became chief justice, helping to bring about a major restructuring of the state court system.

Despite his mandatory retirement at age 70, he continued to hear civil cases on a part-time basis until his 90th birthday. Quoting his grandfather Lucian Baldwin, and perhaps making an unconscious reference to the magnificent bridge that bore his name, he commented "It is a good deal better to wear out than to rust out."

Baldwin died in 1986 at the age of 93, survived by his three sons Lucian, Raymond, Jr., and Tyler. His wife Edith, to whom he was married for 48 years, died in 1970.

The June, 1983, collapse of the Mianus River Bridge in Greenwich sparked a nationwide search for bridges that had used the same potentially dangerous support system which had fallen out of favor a decade and a half earlier. In Connecticut alone, 68 structures were identified as similar in design and would be given priority attention.

CHAPTER 2 Planning a New Bridge

It is ironic that one generation's engineering marvel could become another generation's bottleneck. But that was clearly the case, and it happened amazingly quickly. By the mid 1960s, weekend traffic delays became common again.

Talk of building a companion bridge to handle eastbound traffic began in earnest — that's what was being built over the Thames in New London, why not do the same thing here? This was the mission of Lea Marsh, who spearheaded the group that got the first Baldwin Bridge started. Marsh and his supporters petitioned Hartford once again to build a companion bridge.

This time, however, there was serious local opposition to any new bridge. Fearing destruction of local roads and buildings in historic Lyme, increased traffic, noise and similar issues, a well-funded and well-organized opposition sprang into action, effectively blocking consideration of the proposal to build a new bridge. A lack of funding for the project played an important role, too, and little more was done about new construction until the 1980s.

The need to move ahead with a statewide bridge program was underlined in June of 1983 by the collapse of the I-95 bridge over the Mianus River in Greenwich. Three people lost their lives when a 100-foot-long section suddenly plunged into the river below. The accident was blamed on the failure of a bridge hanger.

A few months later, 519 additional state bridges had been identified by the state as being in "poor" or "fair" condition and requiring immediate attention, with another 1,100 state bridges requiring repair, restoration or replacement over the next 10 years.

In 1984, Governor William A. O'Neill announced plans to initiate and fund a massive 10-year effort to upgrade Connecticut's transportation system. This Transportation Infrastructure Renewal Program was to cost $7 billion and was an unprecedented commitment to rehabilitate and improve the state's highways, roads, bridges, transit operations and other system components.

ENDING THE TOLLS ON THE 1948 BRIDGE

MANY local residents remember the day the tollbooths were torn down. If it hadn't been for Lea Marsh, that day might not have happened when it did.

The tolls were originally put in place as a means of obtaining and retiring the low-interest construction bonds that paid for the bridge. But instead of segregating toll income, receipts ended up in the general fund.

By the time 1968 rolled around, traffic projections made in the 1940s (which indicated it would take 30 years to retire the debt) were greatly exceeded and sufficient receipts had been generated to retire the bonds a decade early.

The state, now accustomed to the added income the 10-cent toll generated, wasn't ready to live without it. Nor was the Connecticut Department of Transportation (ConnDOT), which wanted funds to pay for bridge maintenance. Not only did officials in Hartford plan to keep collecting tolls, but they were still sitting with the $500,000 difference between the $6 million raised and the $5.5 million it actually cost to build the bridge.

Marsh started a one-man crusade to eliminate the tolls. Commuting to Hartford for meetings with the governor, legislators and lawyers, he reminded them that the area's citizens had been promised that tolls would end as soon as the debt was retired. His determination finally prevailed and the tolls were dismantled.

The timing couldn't have been better, as the new Route 9 was nearing completion and the great summer bottleneck was already a problem once again.

The tolls ended on July 4, 1968, and a ceremony with fireworks was held on the bridge to celebrate the occasion. Officials were so engrossed in the festivities that no one noticed an enterprising trucker who reportedly slipped into an abandoned tollbooth and continued to collect tolls from unsuspecting motorists.

The governor's program included funding to repair or replace a number of noteworthy bridges, including:

Mianus River Bridge in Greenwich $66.3 million (reconstruction and widening)
Niantic River Bridge in Niantic - $41.3 million (replacement)
Yankee Doodle Bridge in Norwalk - $13.3 million (rehabilitation)
Quinnipiac River Bridge in New Haven - $25.7 million (rehabilitation)
Arrigoni Bridge in Middletown - $21.7 million (rehabilitation)

While about 46 percent of the funding for those projects would come from state funds (with the balance from federal and other funds), only 10 percent of the funds to rebuild the Baldwin Bridge — the largest project of all — would come from the state. Ninety percent of the funding was already available through previously allocated interstate highway program funds.

Estimates for the entire replacement of the bridge (including the two approaches and demolition of the old bridge) were in the range of $170 million. The bridge itself was expected to cost about $100 million. It would be the largest single contract ever let by the Connecticut Department of Transportation (ConnDOT).

Studying the options

ConnDOT selected the engineering consulting firm of Parsons Brinckerhoff Quade & Douglas, Inc., to perform a

The Parsons Brinckerhoff engineer responsible for the overall design of the new Baldwin Bridge was Czechoslovakian-born Charles Rejcha. He oversaw all the early design work for both steel and concrete alternates. Design work for the superstructure was divided between the firm's Tampa office under Juan Murillo (which designed the concrete alternative) and the firm's Hartford office under Gerald Waltman (which designed the steel alternative). Charles, however, focused his attention on the substructure which was to be used no matter which superstructure option was selected.

feasibility study. Their initial evaluation centered on an attempt to save the existing bridge and use it to carry southbound traffic while a new bridge would be built to carry the northbound lanes. The exising bridge would have to be widened to provide for four lanes of traffic and hard shoulders for breakdown. Also, some of the piers would have to be raised to give the bridge a less abrupt crest at the highest point of the span so that motorists would have a longer sight distance along the roadway as mandated by today's codes.

Added to these considerations, the latest design codes require bridges to be able to support considerably more weight than was called for back in 1948 when the bridge was built. New pier columns would have to be strong enough to support on each of the four lanes and hard shoulders 640 pounds per linear foot or a total of about two tons per linear foot of bridge deck.

The cumulative effect of these requirements would have increased the load on the existing steel piles (which anchor the bridge's foundation to the bedrock and subsoils) by 31 percent or 15 tons per pile. Another concern was that pile driving and construction of the new northbound bridge, even though it would be placed a minimum of 60 feet away, would cause substantial horizontal and vertical movement on the existing bridge due to vibration from construction activities and consolidation of the river bed.

The deck support system, as well as the deck itself, would also require complete rebuilding once the piers were reconstructed. This would provide redundancy in the supporting girders, as is now common practice, so that failure of any single support member would not cause a collapse of the structure such as had occurred in the tragic failure of the Mianus River Bridge. Rehabilitation of the existing bridge was looking less attractive as each new factor was considered.

Nor would it be feasible to keep it just for local traffic or as a pedestrian and cyclist walkway, as had been suggested by a few local residents who felt that this would be very desirable. It would be very difficult to tie in local traffic with through traffic, and the increasing cost of maintaining an aging steel bridge just for local traffic and pedestrians could not be economically justified in the economic climate of the 1990s.

At the end of the study, Parsons Brinckerhoff's evaluation concluded that expansion of the existing bridge was not feasible because of the substantial increase in load on piles and footings (whose condition was unknown), and the high cost of demolishing and reconstructing the existing piers and bridge deck. The cheapest, safest way to relieve traffic congestion was to build a new bridge.

PARSONS BRINCKERHOFF

THE ENGINEERING firm of Parsons Brinckerhoff Quade & Douglas, Inc., was founded in 1886 and is based in New York City. They have established a reputation as a leader in the civil engineering field with hundreds of infrastructure projects including tunnels, dams, superhighways, transit systems and bridges successfully completed.

Among those best known to the public are New York City's first subway (the Lexington Avenue line), New Jersey's Garden State Parkway; Florida's Sunshine Skyway, the Hampton Roads Bridge-Tunnel in Virginia, the NORAD Underground Defense Center in Colorado Springs, the Bay Area Rapid Transit System and Trans-Bay Tube in San Francisco, the Newport Bridge in Rhode Island, the Fort McHenry Tunnel in Baltimore and the Metropolitan Atlanta Rapid Transit System. They also served as engineers for the 1939 New York World's Fair.

Parsons Brinckerhoff has been rated by *Engineering News Record* as the #1 transportation engineering firm for the past four years.

The Sunshine Skyway Bridge in St. Petersburg, Florida, has many similar features to the New Baldwin Bridge except for the twin towers and cable stays. This type of structure was considered during the preliminary design phase and would have served the purpose well. Local citizens, however, wanted a low profile bridge similar in appearance to the old Baldwin Bridge.

1911

1948

1993

These drawings of the 1911, 1948 and 1993 bridges show how engineers have accommodated steadily increasing traffic demands. While the profile of the 1948 and 1993 bridges appears to be quite similar, the cross-section at the right shows the dramatic increase in traffic volume the new bridge can handle.

CHAPTER 3

Design of the New Baldwin Bridge Begins

PARSONS BRINCKERHOFF'S efforts then turned to the design of a new structure. First and foremost was the determination of what type of structure was most suited to the purpose. Many factors were evaluated at this time: design parameters; type of construction; span length; geology; local conditions; topography; environmental restrictions; economics; maintenance of traffic (both on land and on the river); local politics; public concerns; etc., etc.

Nineteen basic structure types of varying span lengths were studied in concept. Of the 19, eight bridge types and spans were selected by ConnDOT for further study. A formal report was prepared that identified bridge types, spans and cost estimates. The report presented all the normal construction options: steel plate girder; steel box girder; precast prestressed concrete box girder; cast-in-place segmental concrete box girder; and precast segmental concrete box girder bridges. Spans varied from 200 to 240, 290 and 480 feet.

ConnDOT selected two types for final design: a steel plate girder bridge and a precast segmental concrete bridge using a 240-foot span arrangement similar to the existing struc-

CONCRETE BOX GIRDER OPTIONS

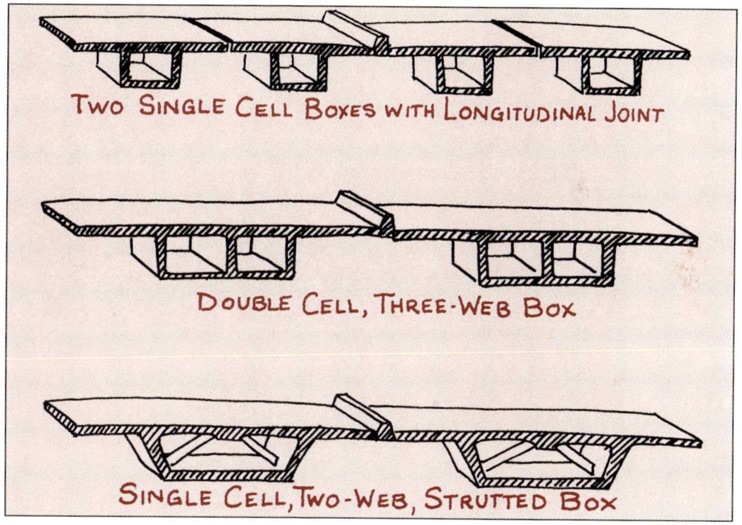

The preliminary design phase considered various types of steel bridges and several concrete design options (shown above). In the end, the single cell, two web, strutted box was selected as being the most efficient, the lightest and the one which would require the least construction time.

Geologic Profile

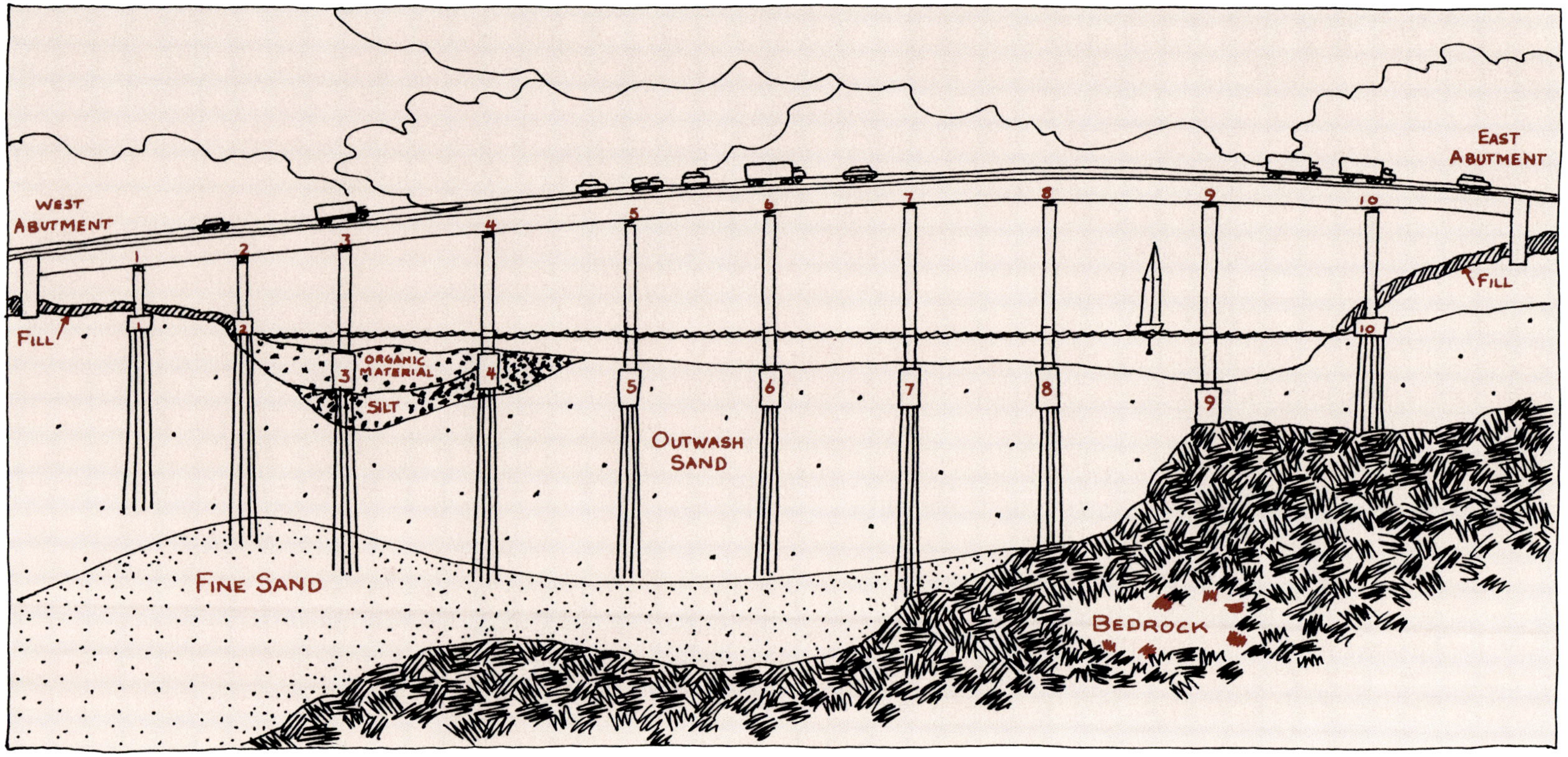

This drawing shows the layout of the two abutments and the 10 piers (numbered 1 to 10 starting on the Old Saybrook side) with respect to the geology of the river bed. The majority of the river bed is comprised of glacial outwash sand and gravels overlying fine sand, and areas of glacial till on top of bedrock. Some organic deposits were identified near the western bank. The bedrock on the west bank is much deeper and the pilings for the pier columns are supported by friction piles. Piers 7, 8 and 10 are supported on piles which extend to the bedrock, while pier 9 is supported on caisson piles drilled into it.

ture. Design work for both would be carried out by Parsons Brinckerhoff in offices scattered across the country in Tampa, New York, Boston and Denver.

It might seem strange to those unfamiliar with this process why it makes sense to go to the expense of designing two bridges when only one design would be used. The Federal Highway Administration, which funds up to 90 percent of interstate projects like the Baldwin Bridge, generally mandates that competitive designs in steel and concrete be considered for all highway bridges in excess of $10 million.

While this adds to the design expense, the competitive environment between design teams usually more than offsets the additional costs. It also gives contractors an option to select which type of construction is best suited to their work force and experience, and may allow them to re-use equipment purchased for a previous project. This policy has also encouraged the steel and concrete industries to achieve greater efficiency in design and construction processes.

That being the case, two design teams went to work, one on a steel bridge, the other on a precast concrete structure. Since the concrete alternative was ultimately erected across the Connecticut, the descriptions of the design work on the following pages relate for the most part, except where otherwise noted, to the concrete bridge.

Geology at the bridge site and foundation design

Critical to any bridge design is the foundation. For a major bridge like the Baldwin, a clear understanding of the local geology was essential. Lying within the southwestern part of the eastern highland of Connecticut, the bridge site is underlain by very old metamorphic rock, chiefly gneiss, schists and granites.

The least erodible rock lies to the east of the river and the deepest bedrock lies more than 250 feet below sea level. The bedrock valley was deepened by glacial erosion and refilled with glacial outwash sediments consisting mainly of sand and gravel. The present rock depth was also affected by sea level change and a downward tilting of the land. Glaciation apparently had little effect on the present course of the Connecticut River.

As a result of this geological history, in which a variety of strata underlie the bridge site. The engineers had to utilize several differing types of foundation support systems, further complicating the foundation design.

Designing the foundations and piers

Parsons Brinckerhoff, working with the New York City-based foundation engineering firm of Mueser Rutledge Consulting Engineers, carefully examined all previous data, including earlier soil borings taken at the site in 1974 and 1984, as well as considerable data collected during the construction of the 1948 bridge.

One of the most important aspects bridge engineers consider is the foundation design. It's almost impossible to have too much data; too little data can ultimately be very costly or worse, catastrophic. The objective was to select the most suitable type of foundation and pile to safely carry the loads, to minimize the possibility of future settlement and to protect against any possible damage from river scour.

To be absolutely sure of subsurface conditions, additional borings were ordered once the number of piers, pier spacing and exact locations were determined. The investigation was carried out in 1988 with test borings at each of the ten piers and two abutments.

A preliminary test pile program was also undertaken. Several types of piles were examined and tested in the preliminary design phase. These included steel H-piles, square prestressed concrete piles and 16-inch-diameter monotube

RIVER SCOUR

Anyone who has ever been to an ocean beach understands the "scour" principle. Stand in the surf and watch a receding wave wash sand away from — and sometimes under — your feet. That's scour.

In a river, the phenomenon occurs when unusually high volumes of water sweep down the river, such as after heavy rains or when the spring snows melt. These currents are capable of lifting and moving river bottom material, particularly sands and silts, from areas where currents are running fast and redepositing them in areas where the current slows. When bridge piers are placed in the water, currents are shifted and serious erosion can result — to the extent that piers can become unstable. Such was the condition at the Schoharie Creek Bridge failure in New York State in 1987.

piles. The square concrete piles or the monotube piles would have been capable of greater individual loads and so a lesser number would have been required at each pier. However, the tests indicated substantial vibration was felt at the pier footings of the existing bridge. The conclusion was reached that a steel H-pile would perform best in terms of efficiency in carrying the design load and, equally important, in creating a minimal disturbance to the existing bridge — only 60 feet to the north.

Even with the steel piles, there was concern that vibration might damage the existing bridge, so a sequence for driving piles was determined, moving from the center to the edge of each cluster and from north to south (away from the existing bridge). It was felt that by so doing, the river sand, which was expected to settle as much as two to three feet, would be less able to transmit vibrations to the old bridge.

To be doubly sure, however, that production pile driving would not cause problems for the aging I-95 bridge when the new bridge construction got under way, special provisions would be included in the construction contract. The contractor would be required to take survey data on the existing bridge at each pier and at each rocker bearing at the end of every day that piles were driven. In reality, over the 12 months that piles were driven, no variation other than seasonal movements was ever detected during production driving operations.

The new piers would also be monitored during construction to detect any settlement which might occur after placement. No significant foundation settlement was observed throughout the construction process. The foundation engineers had done their job well.

The substructure, or pier bases and pier columns, were designed in the Parsons Brinckerhoff Boston office. These are the most heavily reinforced sections in the bridge with rebar (steel reinforcing bars) up to $2\frac{1}{4}$ inches in diameter extending from the pier bases up into the lower column lifts. All rebar in the bridge would be coated with epoxy to increase its useful life. The pier columns are each solid rectangular sections, measuring eight feet by 36 feet, rising to a cap at the top upon which sit the two bridge bearings and an earthquake restrainer.

Originally, earthquake restrainers had been designed for piers 5 through 10 only, but following the Loma Prieta, California, quake in 1989, it was decided to include restrain-

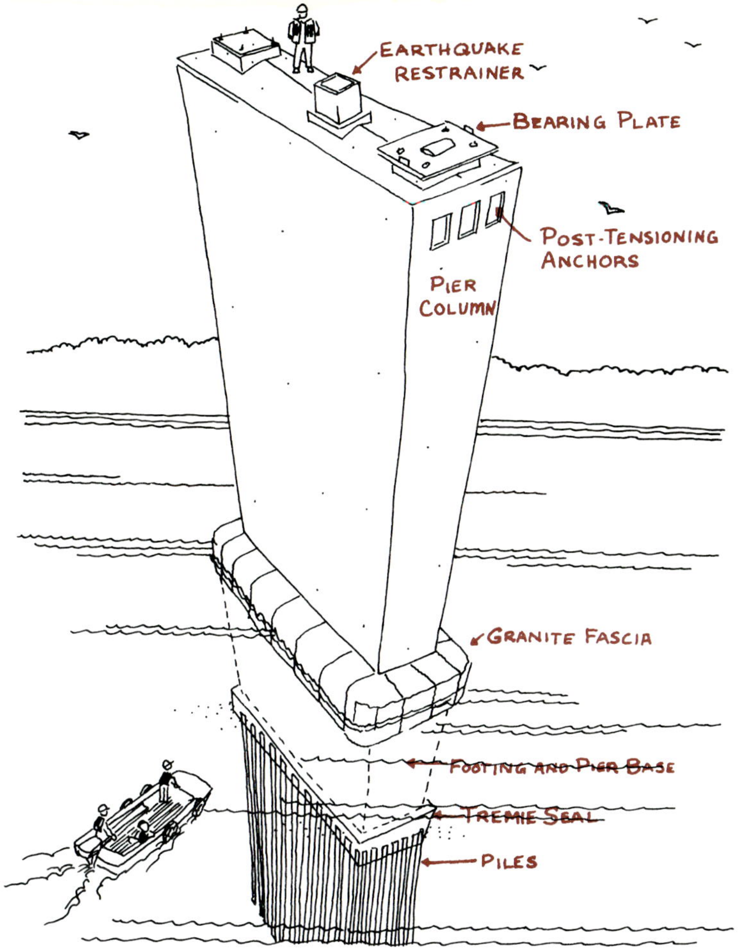

This schematic shows a single pier for the New Baldwin Bridge. At the bottom, steel friction piles carry the loads to strata deep below the river bed. The entire structure, from the tremie seal on the river bottom to the top, is made of reinforced concrete which was poured in place. The top section of the columns is also post-tensioned (see page 43) in both directions to contain the high bursting forces from the weight of the superstructure which sits on the bearing plates. In the center is an earthquake collar which resists sideward movement in the unlikely event of an earthquake.

ers at each pier and each abutment. The purpose of the restrainers, which extend up from the pier cap through a rectangular hole in the pier segment bottom, is to prevent any possibility of the bridge shifting sideways on the pier cap in the event of a major quake.

Designing the bridge deck

With four 12-foot lanes and two 12-foot hard shoulders in each direction, the roadway design required the eastbound bridge to have a deck width of 76 feet, 11 inches. The westbound bridge, which also has an eight-foot-wide pedestrian and bicycle path, measures 86 feet.

This creates a 163-foot deck width, which had few precedents in a segmental concrete structure and resulted in an unusual cross-sectional layout. Conceptual studies analyzed all options, and a single-cell box with internal struts was selected for final design. Sloping webs produced favorable balance between interior and exterior cantilever slab spans while limiting pier widths to a minimum dimension and improving aesthetics.

With the exception of the internal struts on the typical segments, all elements of the cross section are post-tensioned both longitudinally and transversely (lengthwise and crosswise). Post-tensioning inhibits cracking and results in a more durable structure.

As part of recent requirements for maintenance of interstate bridges, the design had to provide for total deck-slab replacement, with the exception of the spine beam and the deck slab portion above the two webs. This had a significant impact on overall cross-section design because tendon

CONNECTICUT EARTHQUAKES

WHILE some might consider earthquake restrainers on a Connecticut bridge a needless expense, the fact is that considerable seismic activity (in frequency, if not in magnitude) has been recorded here. The center of activity in the Connecticut River Valley appears to be the East Hampton-Moodus area, where a quake measuring 2.8 on the Richter scale was felt as recently as March 11, 1992. Seismologists estimate that 100 or so minor tremors occur in our state every year, but most are so small they go unnoticed by the average person.

Although the mountains and valleys of New England are crisscrossed with fault lines, they luckily do not have a history of generating much seismic activity. Nevertheless, an earthquake of magnitude 3.5 or so strikes some part of Connecticut every five or six years. And the reverberations of a 6.0 earthquake, centered 90 miles north of Quebec, were felt in our area in November of 1988.

So earthquake restrainers on the Baldwin Bridge are a necessary precaution, even if the likelihood that they will be needed is quite slim. Its design provides that approximately 20 percent of the weight of the superstructure must be resisted in the horizontal direction by the restrainers [20% x 78,623 tons (design dead load) = 15,725 tons].

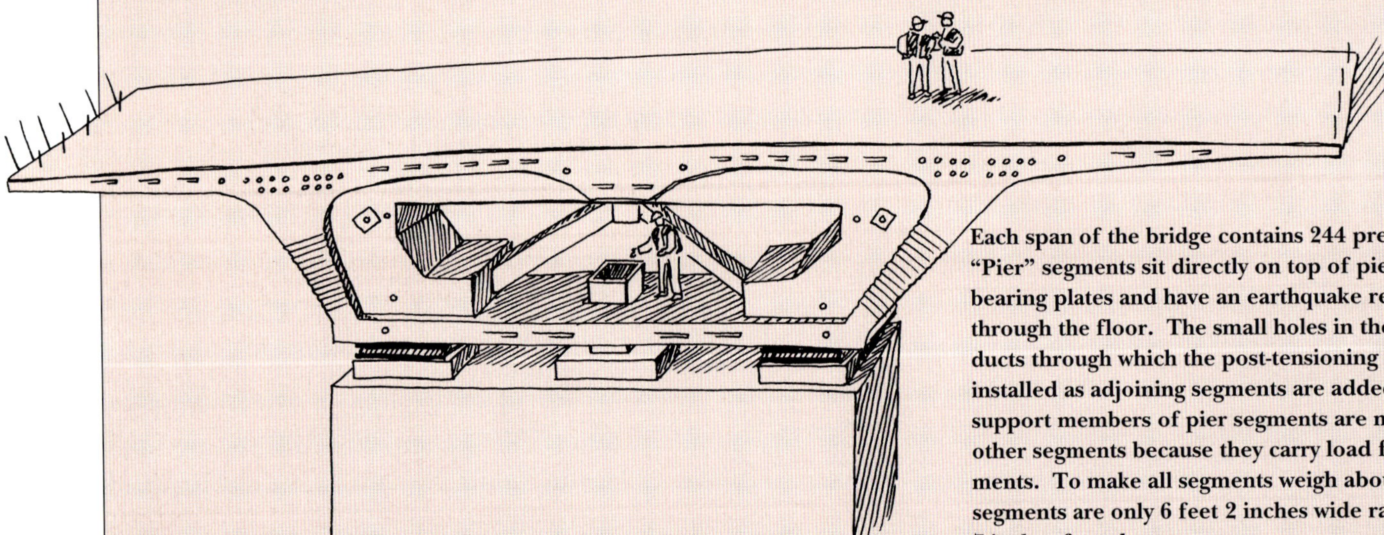

Each span of the bridge contains 244 precast segments. "Pier" segments sit directly on top of piers. They rest on bearing plates and have an earthquake restrainer protruding through the floor. The small holes in the top deck are the ducts through which the post-tensioning tendons will be installed as adjoining segments are added. The walls and support members of pier segments are much thicker than other segments because they carry load from adjoining segments. To make all segments weigh about the same, pier segments are only 6 feet 2 inches wide rather than 10 feet 7 inches for other segments.

POST-TENSIONED CONCRETE

CONCRETE is a superb building material. What makes it particularly useful is that it is delivered to the site as a liquid and can be "poured" into forms of virtually any shape. Most concrete is poured into forms containing steel rods called reinforcing bars (or rebar) which provide extra strength. This is commonly referred to as reinforced concrete.

While concrete is most efficient when it is used to resist forces of compression, steel performs exceptionally well when it is stretched (in a state of tension). Post-tensioning takes advantage of the inherent qualities of both materials and was employed on the bridge to give it additional strength.

In a post-tensioned structure, the concrete is poured into the customary forms, but in addition to rebar, the form contains ducts through which high-strength threaded bars or wire strands are installed. After the concrete cures, but before any loading is placed on it, the steel running through the concrete is stretched by hydraulic jacks. Gripping wedges or threaded nuts clamp the stretched steel to the concrete, locking in the compressive forces forever.

Any future loads which try to stretch or put tension into the structure will be resisted by the post-tensioning forces and the structure will remain in compression at all times.

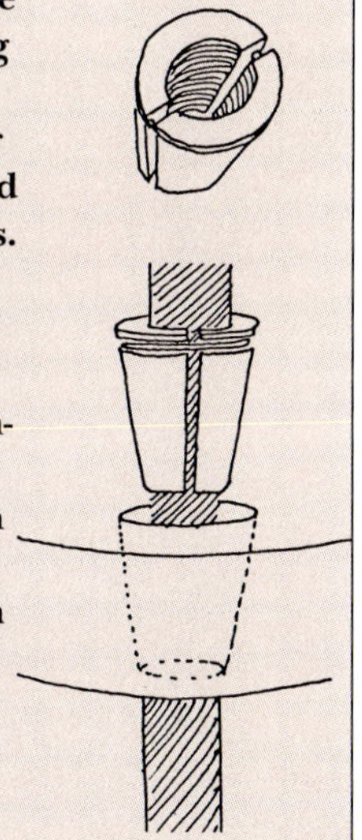

The post-tensioning process requires hardware to lock tensioning forces into the wire strands after they are tensioned by hydraulic jacks. The two-part wedge is tapered on the outside to fit tightly in a conical hole in the anchor plate. On the inside are sharp teeth which tightly grip the strand as it is squeezed back into the anchor plate.

ducts and anchor plates had to be placed a minimum of eight inches below the top of the deck to allow for eventual resurfacing.

Those that have visited the Tampa/St. Petersburg area might recognize certain similarities between the Baldwin Bridge and a somewhat similar bridge over Tampa Bay — the Sunshine Skyway. Many of the features of the New Baldwin Bridge were also successfully used in the Skyway Bridge which was opened to traffic in 1987. While the central portion of the Skyway Bridge is supported by cables from two towers, the remaining high-level portion, with its 240-foot spans, is remarkably similar to the Baldwin.

A FIRST FOR CONNECTICUT

PRECAST concrete segmental bridges were pioneered in Europe in the 1950s; by the 1970s the technology had spread to the United States. The bridge across the Connecticut, however, would become the state's first of this type.

The first precast concrete segmental bridge in this country was the J.F.K. Memorial Causeway in Corpus Christi, Texas, which opened to traffic in 1973. Today there are probably more than a hundred precast concrete segmental bridges in the U.S. either completed or under construction.

What's unusual is that the design of the 2,626.5-foot-long Baldwin Bridge calls for cross-sections that are among the heaviest and widest used in the country for cantilever construction, surpassed only by the segments of the Sunshine Skyway Bridge in Tampa, Florida.

In November, 1990, the car barge extended from the Old Saybrook shore as the construction of cofferdams and piers progressively moved across the river from west to east. Work has already begun on the Old Lyme shore digging footings for pier 10.

CHAPTER 4 Selecting A Contractor

THE BALDWIN BRIDGE contract was advertised to prospective bidders in the fall of 1989 and the bids were opened on December 13, 1989.

Eleven bids were received from contractors and joint ventures of firms proposing to construct the new bridge. Interestingly, 10 of them involved building a steel bridge, while only one bid proposed the concrete method. The single concrete bid, however, from a joint venture of Perini, PCL and O&G Industries (referred to by the initials of the three firms, PPO) was the low bid at $93.8 million. The lowest bid for the steel alternate was for $99.98 million.

Ted Kirk, PPO assistant project manager, explains why they chose concrete. "It all boiled down to lower cost and greater control," he said. "Not only are the material costs of a concrete bridge lower, but with a steel structure, the contractor must rely on the schedule and workmanship of both an off-site fabricator and a subcontract steel erector. We decided that the concrete segmental alternate would allow us to do all of the work on site with our own forces giving us complete hands-on control over both the workmanship and schedule."

The Perini/PCL/O&G (PPO) joint venture was awarded the contract in January of 1990 and official notice to proceed with the work was given on February 1, 1990.

Three partners; three specialties

PERINI

Perini Corporation, one of the oldest, largest and most diversified contractors in the United States, is based in Framingham, Massachusetts. It will celebrate its 100th anniversary in 1994. Among Perini's impressive highway and bridge projects is the Massachusetts Turnpike Extension, which connected the turnpike with downtown Boston. In 1965, this was the largest single highway contract ever awarded in the United States. Perini has also been responsible for a wide range of construction in other areas including transit systems, power plants, pipelines, waste water treatment plants, airport facilities and all types of building construction both here and abroad.

Recent projects include Boston's Charlestown interchange and related highway bridge and tunnel construction, a portion of the Boston Harbor cleanup, mass transit reconstruction in New York and the massive Metro Redline Transit System in Los Angeles.

Perini was the lead company in the joint venture for the Baldwin Bridge and had overall responsibility for satisfactory completion of the work.

PCL

PCL Constructors, Inc., based in Edmonton, Canada, is the largest civil contractor in Canada. They brought their expertise in segmental concrete bridge construction to the joint venture. PCL had already earned a distinguished reputation for successful bridge construction in this field in Canada and the U.S., having completed bridges similar to the Baldwin, including several as part of the Port Everglades Expressway in Florida, the rapid transit bridge and James McDonald Bridge in Edmonton, the Glade Creek Bridge in West Virginia and an impressive cable-stayed bridge over the Fraser River at Annacis Island near Vancouver.

PCL's input at the casting yard and later in the segment erection operations would be a key factor in the overall success of the project.

O&G

O&G Industries, Inc. of Torrington, Connecticut, is a family-run general contractor firm with over 60 years' experience in the private and public sectors. Projects constructed by O&G over the years include many hundreds of miles of state highway, including sections of Interstate I-84 and I-91, and the construction of water treatment facilities, dams, buildings and bridges. O&G also has a strong materials division in Connecticut with seven concrete plants, 10 asphalt plants, four quarries and four sand and gravel plants.

O&G'S contribution to the joint venture would be based mainly on their familiarity with work in Connecticut, obtaining qualified subcontractors and high-quality vendors, etc. Also significant was their expertise in the field of dispute resolution, an area spearheaded by Raymond Oneglia, Jr.

There was understandable apprehension within ConnDOT when the concrete design won the bidding because this would be the first concrete segmental bridge built in Connecticut. In fact, there were only two segmental bridges in New England — one in Wiscasset, Maine, completed in 1982 and the second at Jamestown, Rhode Island, started in 1986 and, at the time, still at least two years away from completion.

Dr. Robert Gubala, former chief engineer for ConnDOT, had visited the Wiscasset Bridge a year or two earlier. It was much lower and closer to the salt water than the Baldwin Bridge would be and was therefore exposed to far more severe conditions. The bridge, which had been in service for seven years, appeared to be in very good condition and the Maine DOT was very pleased with its performance. Any reservations he had held about segmental bridges were allayed.

PCL, with their extensive experience in building concrete segmental bridges in Canada and the U.S., provided the expertise that allowed the PPO joint venture to put forward a plan for completing the bridge in three years rather than the four years allowed for project completion by ConnDOT. Such an aggressive schedule would require close coordination of all the suppliers and subcontractors involved and also a close working relationship with ConnDOT and the engineering consultant brought in by the department to administer and supervise construction.

For the consultant supervision, ConnDOT elected to use the construction branch of Parsons Brinckerhoff, Parsons Brinckerhoff Construction Services, Inc., which had considerable experience in this field and had also been the lead firm in a joint venture consultant team that had supervised the Sunshine Skyway Bridge.

Early site meetings set the tone for a team effort by all parties, with commitments made to address problems as they arose, resolve them and get on with the bridge. This team approach created a means of solving disagreements quickly and proved to be a major reason why the bridge would be opened to traffic more than nine months ahead of schedule.

The Parsons Brinckerhoff team was augmented by inspectors and field survey personnel of L-C Associates, Rocky Hill, Connecticut. Under the direction of company president Dr. Frank S. Chuang, L-C has been involved in the design and inspection of many infrastructure projects in Connecticut and New York state and is well known to ConnDOT.

BIDDING ON CONTRACTS

To open up the bidding and allow smaller contractors the opportunity to participate, the construction was to be split into four contracts. The bridge was bid in December of 1989. Each of the two approaches was bid separately in 1990. The demolition will be advertised in summer of 1993 and is expected to begin in the fall.

1. **Construction of the New Baldwin Bridge**
   Contractor - Perini/PCL/O&G joint venture
   Contract start - February 1, 1990
   Contract value - $93.8 million

2. **West approach**
   Contractor - The Middlesex Corporation
   Contract start - July 30, 1990
   Contract value - $14.4 million

3. **East approach**
   Contractor - Manafort Brothers, Inc.
   Contract Start - September 19, 1990
   Contract value - $7.8 million

4. **Demolition of the old bridge and construction of the new boat launch area**
   Contractor - yet to be determined
   Expected contract duration - approximately 18 months

Changes in construction plans

PPO then submitted its three-year target schedule and requested ConnDOT's support. This was critical since the plan also involved approval of a barge system beyond what was provided for in the permit — extending barges two-thirds of the way across the river so materials and equipment could be delivered where they would be used.

Since the building permits were issued to ConnDOT, it was necessary to request changes to the permits to allow this extension. This was accomplished through a series of meetings with all the permit agencies involved in the construction project, including ConnDOT's own environmental group, the Department of Environmental Protection and the Coast Guard.

The first two groups were concerned about the environmental aspects of the work and ensuring local boating safety. The Coast Guard, of course, had to be sure that the effects on recreational and commercial river traffic would be acceptable and would not create any unnecessary hazards to navigation.

After several sessions with these agencies, approval was finally given. The revised permit provided that a secondary channel for recreational boaters be established between piers 6 and 7 during the peak boating months between May and early November. It would remain open at all times during this period except for major concrete placements which had to be scheduled for mid-week. A series of warning signs and navigational lights were also worked out with the Coast Guard. River traffic was kept updated of changing configurations through weekly Notice to Mariners bulletins, issued by the Coast Guard. Local boaters were kept informed by the posting of a notice board at the adjacent boat slip and by flyers issued to local marinas.

Another major change PPO proposed involved the method by which the precast bridge segments would be put in place. This differed from the erection scheme shown on the contract drawings, which suggested using the beam and winch system to lift segments from the ground or from barges in the river.

The contractor and its engineering consultant, Finley McNary Engineers, proposed an erection method which used a moveable gantry (called a launching gantry because of its ability to launch itself forward as work progresses and continually be able to place bridge segments in their proper location).

This method had previously been used with great success both overseas and in the U.S., most recently in Colorado's Glenwood Canyon where Finley McNary was also the construction engineer. PPO engineers went to Colorado to see for themselves how that operation was going and to help them decide if they should follow suit. They were impressed by what they saw and by contractor Flatiron/Prescon's satisfaction with the gantry scheme as well as Finley McNary's engineering work.

Elie Homsi, the contractor's engineer responsible for the segment erection, talks about the gantry scheme: "If we didn't use a launching gantry, we'd have needed two schemes for erection — one for segments over water and another for those over land. We'd have spent a lot of time moving equipment to keep that navigation channel open, especially because during major holiday weekends (plus the day before and after the weekend), we weren't allowed to work at all.

"Segments would have to arrive by barge and we'd have to deal with the current — it's pretty swift here. Maneuvering barges into a stationary position so cranes could lift segments would be real tricky. Add in the size of those segments and the 85-foot height we'd be lifting them, and you're talking about big cranes and a lot of barge and tug activity in the water. If there was an alternative to working from the water, that's the one we wanted to use."

Although the capital cost of the launching gantry equipment was greater, the higher costs would be offset by a much faster speed of erection, an earlier project completion and considerable savings in reduced overhead. The launch gantry also offered a single, uncomplicated method for segment delivery and installation. It eliminated the need for segments to be loaded onto barges and floated to the site — a costly and difficult process at best especially with typical river currents on the ebbtide of about $3\frac{1}{2}$ miles per hour. A situation of heavy ice flows such as was experienced during the cold winter of 1991-1992 would also have shut down operations completely if the beam and winch system had been used.

ConnDOT liked the concept and, after a detailed review of the engineering involved, gave the necessary approvals. The stage was set. Now it was up to the contractor to show what it could do.

Bob Novynka and Elie Homsi, erection superintendent and erection engineer for PPO. Both men were heavily involved in the set-up and operation of the casting yard until the summer of 1991 when their focus switched to segment erection.

Segment erection in Glenwood Canyon, Colorado, using a launching gantry similar in concept to the one used on the Baldwin Bridge. PPO was impressed by the progress of the work in Colorado and chose to propose a launching gantry as an alternate to the beam and winch system.

The old bridge provided an excellent vantage point to view the progress of pier construction as work moved across the river to the Old Lyme side. This photo in late 1990 gives some idea of just how much equipment was involved in this phase of the work.

CHAPTER 5 The Start of Construction

THE FIRST ITEMS of work in February and March of 1990 involved clearing the site on both sides of the river so construction in those areas could begin. Several buildings had to be demolished along with the Yacht Distributors Marina and boat slips.

To allow construction barges access to the river banks, joint venture subcontractor Gates and Whaling City dredged the areas on both sides to a depth of 15 feet. In fact, the dredging work had to be started almost immediately because an environmental restriction in the contract prohibited dredging after April 1st due to the annual upstream migration of the Connecticut River shad which typically begins in mid-April. As a further environmental consideration, restrictions on pile driving operations were limited to one or two piers at a time so that noise and vibration in the water would not discourage the shad from making their annual upstream pilgrimage.

Test pile program

When the pier locations were cleared and accessible, a series of tests were conducted at each pier to confirm the load-carrying capacity of the piles and to establish final lengths so the piles at each pier could be ordered.

Each of the 10 piers carry their loads way below the concrete foundation by means of steel piles. The design for each pier was different and the number of piles at each pier varied from 136 at pier 1 to a maximum of 320 at pier 6.

Testing is performed by driving piles (typically 20 feet longer than the length that is actually anticipated) to monitor soil conditions at and below the anticipated depths and by counting the number of hammer blows per foot required to install the pile, called the driving resistance. The soils engineers can equate these field observations to the load capacity of the pile from data collected over the many years and correlated into empirical formulas and charts.

At selected piers, an additional load test was also performed on piles which had already been driven to the design depth. This called for a large hydraulic jack to apply load in increments until $\frac{1}{4}$ inch of movement was detected. For the test, the load at which this movement takes place was considered the capacity of the pile.

The piles at piers 1 through 6 develop their capacity through skin friction with the sand through which they pass. These piles are called friction piles and have a design capacity of 100 tons working load. The piles at piers 7, 8 and 10 carry their load all the way to the rock surface. These piles are classified as end-bearing piles and can carry a significant-

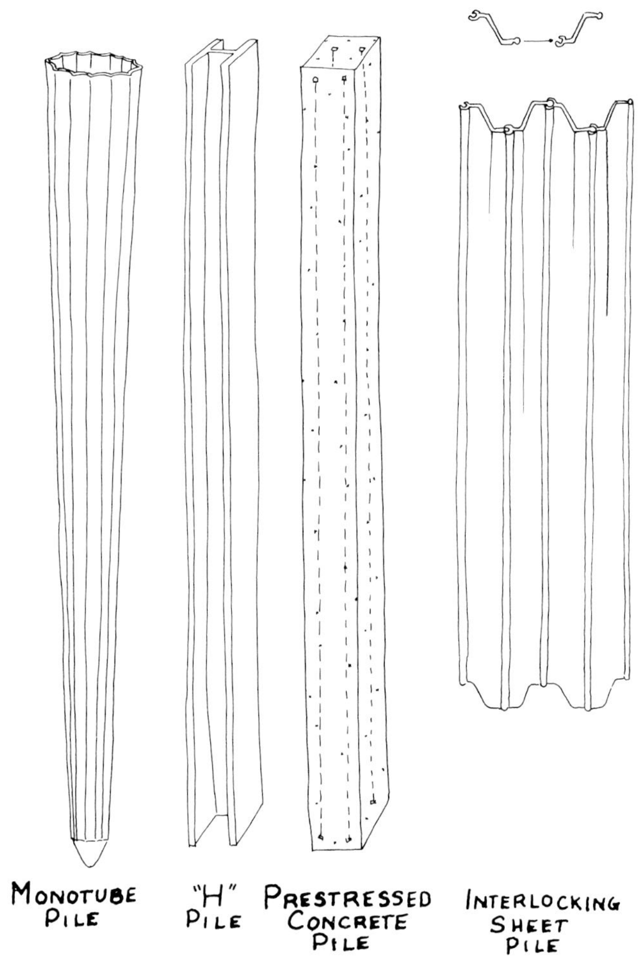

These are the types of piles typically found in bridge construction. The load-bearing piles at the Baldwin were H piles, selected because of concerns about vibration and possible damage to the aging existing bridge. The interlocking sheet piles were used for the cofferdam construction.

ly greater load up to 165 tons. Pier 9 was unique and utilized yet another type of support system which is described on page 59.

Construction of the cofferdams

Once the actual pile lengths were established, the orders were placed with Bethlehem Steel Corporation of Bethlehem, Pennsylvania. It would be eight to ten weeks before the piles would arrive on the bridge site. In the interim, construction of the cofferdams and excavation of the river bottom inside them could begin.

A visitor to the site in late spring or early summer of the first year would have seen a large assembly of work barges, car barges, cranes and other marine equipment necessary for starting cofferdam construction in the river. All of the substructure work was under the direct supervision of Maurice Berry, the general superintendent of PPO.

Eight cofferdams had to be built for piers 2 through 9. Piers 1 and 10 on either side of the river were situated above water level so that cofferdams were not required.

Following cofferdam installation, subcontractor Sea-Jay Marine excavated the river bottom using a clambucket mounted on a work barge and with high-pressure water jets and airlifts around the sheet pile perimeter to remove material not accessible to the clambucket. Final excavation would be confirmed by divers, and soundings would be taken by the inspection team across the entire cofferdam area.

All river bottom material removed from the cofferdams was placed in a special barge with doors in the bottom, called a "dump scow." Each time the barge was filled, it

HOW A COFFERDAM WORKS

A COFFERDAM is a temporary dam or shell made up of interlocking sheet piles formed into a closed loop, like a box with no lid; the bottom is the river bed. Once the loop is completed, excavation of the inside area can be done without surrounding river bed material slumping into it. After the bearing piles are driven and the bottom sealed off, water can be pumped out and construction of the pier can be performed below water level "in the dry."

The sheet piles typically extend about 20 to 25 feet into the river bottom and are braced at several elevations by steel frames to resist the pressure of the outside water when the water inside the cofferdam is pumped out. The sheet piles are often pulled out and reused once the pier is complete. The Baldwin Bridge design, however, called for the sheet piles to remain in place and to be cut off six inches above the top of the reinforced footing. The purpose of having the lower section in place is to provide long-term protection to the foundation, seals and piles against scour and erosion. The bottom of the sheet piles are set at the level which design engineers have calculated as the possible depth of scour from the intensity of a once-in-a-100-year river flood.

Vibratory hammer model ICE 812 installing cofferdam sheeting at pier 8. The west temporary fender is seen in the background.

After the cofferdam is constructed, the next step is to excavate the river bottom material inside. Sea-Jay Marine excavates river bottom by clambucket and deposits it into the scow barge.

Early pile driving operations at pier 3. In this photo, 12 piles are lined up in a template to ensure accurate placement. A vibratory hammer will drive the piles to the template level. The template is then removed and final driving is performed using an underwater hammer.

Parsons Brinckerhoff inspector Tim Cooper watches the Dutch IHC underwater hammer drive the pile to final tip elevation. He records the driving resistance (number of blows) for each foot of penetration. Driving resistance which fell below the criteria of 25 blows per foot sometimes resulted in driving to greater depths or in installing additional piles.

would be towed to an artificial reef in Long Island Sound 15 miles off New Haven and its load dropped into the sound. An inspector from the U.S. Corps of Engineers accompanied each barge to ensure environmental compliance.

Pile driving begins

On schedule, the load-bearing piles arrived at the site. They were shipped by rail from the Pennsylvania mill to a marine terminal in New London and then barged to the job site. Pile lengths varied from the shortest of 60 feet (at pier 10) up to the longest of 135 feet (at pier 7). All piles longer than 120 feet had to be spliced from two shorter sections. Splicing was completed on Admiralty Pier in New London by welding two pile lengths together using a special steel splice sleeve at the joint welded to both pile ends.

Then came the noisy part — driving the piles. In all, there were 1,755 piles to be driven; almost 17 million pounds of steel. Most of the driving was done by two types of pile hammers, and actually, both were surprisingly quiet for pile driving hammers.

The piles were installed in groups by sliding them through a template which precisely located each one. An ICE 812 vibratory hammer was first used to drive, or more accurately to vibrate the pile through the upper strata and take the top of the pile down to the template level. This type of hammer works very well in the sandy materials mostly found in the upper layer of the river bottom.

After removal of the template, all final driving from that point down was done by a Dutch IHC S-70 Underwater Hydro Hammer — a relatively new and innovative device capable of driving the piles well below the water level to their proper elevation. The hammer was equipped with an electronic sensor which could record the variable energy level being delivered to the pile, the penetration and blow count. Manual counts were always made by inspectors to verify driving resistance. Any blow count of less than the 25-blows-per-foot criteria resulted in driving to deeper depths, or in some cases, addition of extra piles. These instances were fortunately quite few. The IHC hammer was also notable for being an exceptionally clean piece of equipment — a big plus in these days of environmental awareness.

The only pier at which blow counts were consistently less than anticipated was at pier 5, where final blow counts as low as 15 blows per foot were recorded. Mueser Rutledge was called back in to review the field data for ConnDOT. PPO sought expertise from their geotechnical consultant Goldberg Zoino (which had earlier provided PPO with data regarding soil conditions that aided in the selection of pile driving equipment).

In a meeting exemplifying the cooperative spirit of all parties, it was decided to perform dynamic testing of three piles at pier 5. In a dynamic test, electronic instrumentation is attached to the pile being driven which detects the magnitude of the hammer impact and the reflection of the impact wave coming back from the pile tip. The tests showed to the satisfaction of ConnDOT that the required capacity of 100 tons per pile had been achieved; the pile crew moved on to the next pier.

The next step is pumping out, or dewatering, the cofferdam. But before pumping could begin, a concrete seal (known as a "tremie" seal because of the tremie pipe

After the piles are driven and the tremie is placed, the cofferdam can be dewatered. This is pier 3 which was the first cofferdam constructed. The tremie was placed August 21, 1990, and the cofferdam dewatered a few days later.

through which concrete is poured to reach the river bottom) had to be placed underwater to seal the bottom. The seals are eight to ten feet thick; each placement involves depositing around 2,000 cubic yards of concrete.

The seal, as the name implies, seals off the bottom of the excavated footing to prevent water from flowing through the sandy bottom and into the cofferdam. Just as important, the seal also acts as weight to hold the dewatered cofferdam down and resist forces tending to float it. This can be likened to a sunken boat, which can be floated when the water inside is pumped out as part of a salvage operation. Here, however, the object was to avoid any such excitement. In reality, the piles bond with the tremie and also act to hold down the cofferdam and the seal by carrying the uplift loads deep into the sandy bottom.

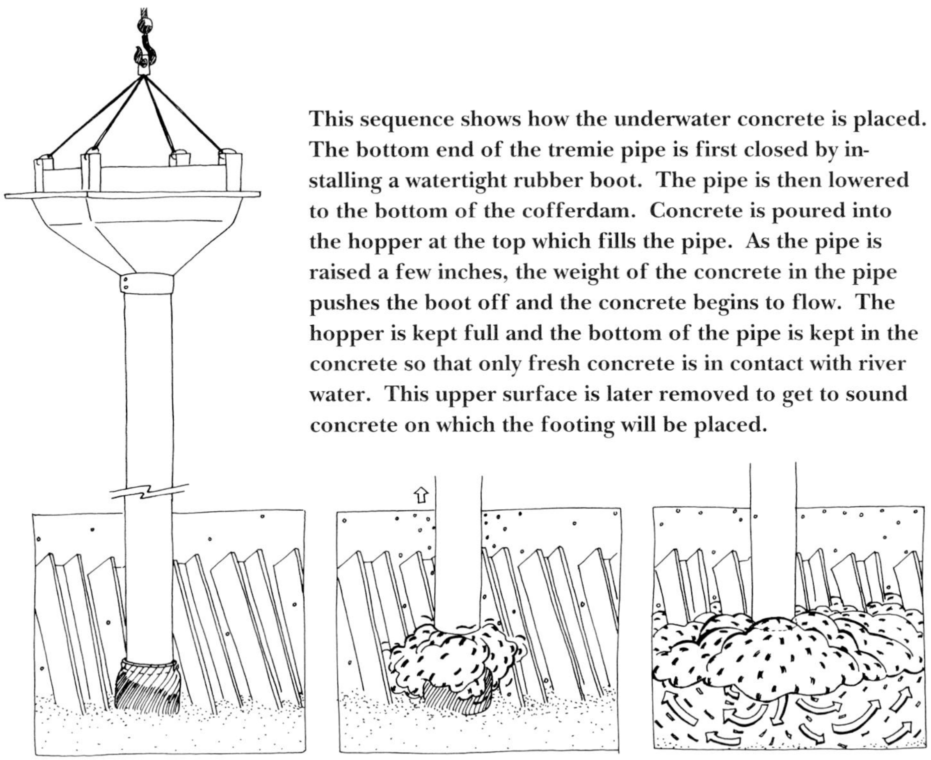

This sequence shows how the underwater concrete is placed. The bottom end of the tremie pipe is first closed by installing a watertight rubber boot. The pipe is then lowered to the bottom of the cofferdam. Concrete is poured into the hopper at the top which fills the pipe. As the pipe is raised a few inches, the weight of the concrete in the pipe pushes the boot off and the concrete begins to flow. The hopper is kept full and the bottom of the pipe is kept in the concrete so that only fresh concrete is in contact with river water. This upper surface is later removed to get to sound concrete on which the footing will be placed.

Once the tremie concrete was placed and the concrete had cured sufficiently — usually about two days — the cofferdam could be dewatered. This typically involved at least two eight-inch submersible pumps, pumping about 5,400 gallons per minute. The actual volume of water to be pumped out considerably exceeds the approximately 1.25 million gallons inside due to leaks in the interlocking seams of the sheet piles. These leaks eventually become sealed naturally by silting and artificially by dropping fine cinders over the outside edge of the cofferdam which are sucked into the seams and stop the flow.

Typically, it took about one full day to pump each cofferdam. The water was pumped directly into the river until the floating pumps reached within a few feet of the bottom. The discharge water was then pumped through a filtration system before it was discharged into the river. Turbidity levels in the river were monitored daily by ConnDOT environmental inspector Milton Beckwith, who was on the project full time until the construction on the river was completed in the spring of 1992.

Once the cofferdam was dewatered, a major cleanup operation went into action to remove the laitance, a cementitous mud, from the top of the tremie concrete using shovels, water jets and vacuum trucks. At the same time, the survey crews established line and grade and marked the

SEQUENCE OF PIER CONSTRUCTION

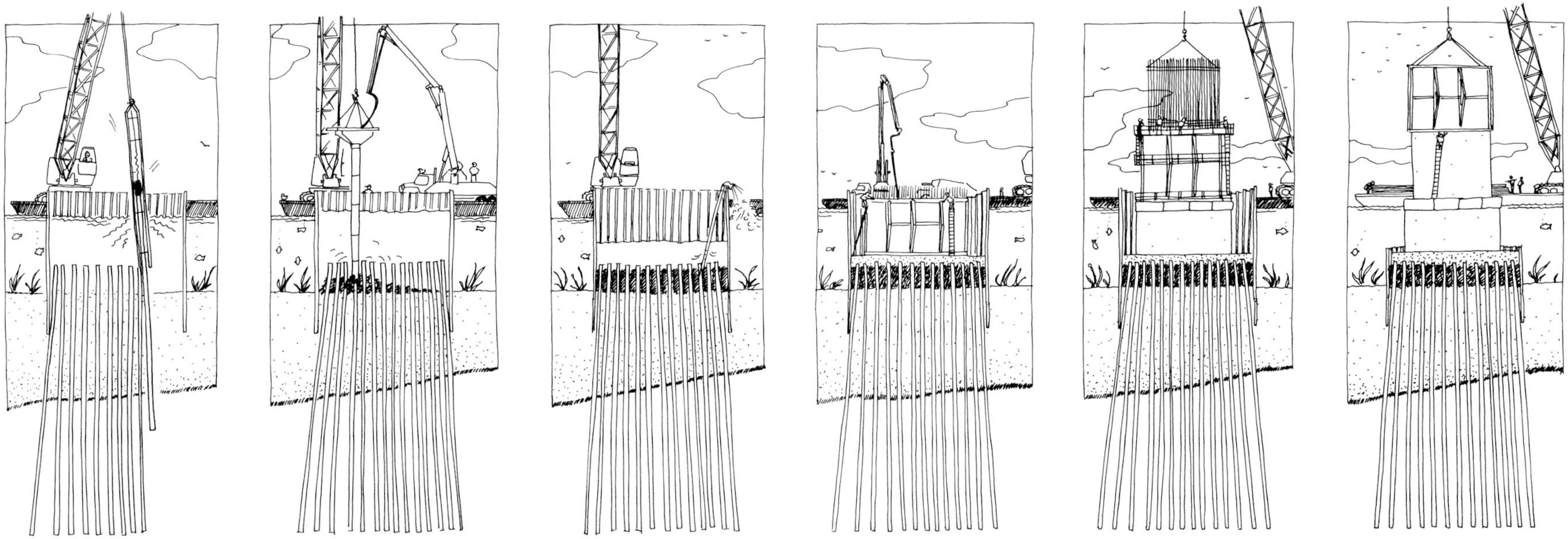

1. Cofferdam driven, excavation complete, driving piles
2. Placing tremie seal
3. Cofferdam dewatered
4. Footing placed, construction of pier base underway
5. Pier base and granite lifts in place; rebar being tied in
6. Form panels raised by barge-mounted crane to next "lift"

theoretical top of the driven piles. They would then be cut off at a uniform level to extend 12 inches upward into the steel-reinforced foundation base.

Then a rebar crew of about 8 to 10 ironworkers from subcontractor Advance Steel Reinforcing Company moved in to tie the reinforcing steel cage for the pier footing in place prior to concrete placement. Advance Steel had the largest subcontract on the project, with about 12 percent of the total contract value. It is headed by President Theresa DiCocco. Advance Steel's rebar superintendent Bill Purcell, who was in charge of all the substructure rebar placement, deservedly won the praise of William Sasser, ConnDOT project engineer, who said "the rebar work was probably the best I have ever seen in all my years with the department — they really did an excellent job."

Some of the pier footings were reinforced with over 300,000 pounds of rebar. It would take an average of 10 days to tie the footing rebar cage in place and to get ready for the concrete. Pours of about 100 cubic yards per hour usually began early in the morning and typically ran into two shifts and up to 24 hours. As many as a dozen concrete transit mixer trucks would be active during these placements, with

The prefabricated bracing frame for the cofferdam at pier 9 was lowered into position by two Manitowoc 4100 cranes on Valentine's Day, 1991. The frame was constructed in another location and floated to the site.

the drivers generally working staggered 12-hour shifts (and by law, not more than 14 hours in a 24-hour period).

Although the substructure work progressed through the winter of 1990-1991, winter temperatures were never really a problem — at least not for the concrete. The concrete plant uses hot water in the mix from mid-November to mid-March to ensure that concrete temperatures at delivery are in the range of 60 to 65 degrees Fahrenheit. However, due to the large concrete volumes in the pier placements, tremendous heat was generated from the hydration of the concrete as it hardened, often reaching up to 160 degrees after two or three days. While the cold temperatures never compromised the strength of the concrete, it was little consolation to field crews working through those cold winter nights when wind chill often made it feel well below zero.

Problems on pier 9

One foundation was quite a bit different from the rest. Look back at the geologic profile on page 38 and you'll see that the bedrock comes much closer to the surface on the east shore. At the location of pier 9, not only does the rock rise close to the surface, but the depth of overlying river sand and silt is also at a minimum — as little as eight feet at the south end.

Not clear from the diagram, but definitely further complicating this scenario, is the fact that the rock surface actually slopes not only from north to south but also from east to west, creating a very difficult base on which to found a bridge pier. The initial solution developed by Parsons Brinckerhoff was to call for excavation inside the cofferdam all the way to the rock surface, upon which the tremie seal would rest. The bedrock was assumed to be capable of carrying at least eight tons per square foot in bearing. The slope load would have been carried by 12 shear pins into the rock.

ConnDOT, however, had recently experienced problems with a similar type of foundation at the Niantic River Bridge, just a few miles up the coast, which had required costly repairs. They decided they would no longer rely on what's called a "structural tremie," where the structural load is carried directly through the tremie seal to the rock below. Parsons Brinckerhoff was asked to modify the foundation design to provide full vertical load through the tremie to the bedrock. This was done by sinking 28 shafts or caissons, each 30 inches in diameter, into the bedrock. In effect, what resulted was compared to a "belt and suspender design" where the benefit of both is achieved.

At the Baldwin's pier 9, each caisson was a drilled shaft going down at least 21 feet into the bedrock. Then a heavy tubular pile section — approximately six times heavier than the typical piles at other piers, was lowered into the hole and sealed with high-strength cement grout to permanently fix it in place. These heavy pile sections then extend up through the tremie concrete seal into the foundation and are capable of carrying the vertical loads typical to other piers and, in addition, resisting the tendency to drift down the ledge. The drilling operations were carried out by Rhode Island-based subcontractor A & W Drilling.

Pier 9, in fact, proved to be the greatest challenge not only for the designer but also for the contractor and took by far the longest time to construct. Unlike the constructed-in-

PPO field engineer Bill Lovely stands beside one of the rotary drill bits used to drill the 30" diameter holes into bedrock at pier 9.

Looking down on the drilling operations at pier 9 shows some of the complexity of work at this pier, which proved to be the biggest challenge of all the substructure work.

PNEUMATIC CAISSONS

THE TERM caisson used here is quite different from the type of pneumatic, or compressed air, caisson that was used for large pier foundations in the late 19th and early 20th centuries. Notable structures still in existence today that used such caissons include James Ead's St. Louis Bridge (1874), John Roebling's Brooklyn Bridge (1883) and Benjamin Baker's Forth Bridge in Edinburgh, Scotland (1890).

In those days, a very large watertight box would be constructed of wrought iron or timber. It would have a heavy roof and strong sides with iron cutting edges at the lower perimeter. This would be floated to the site and sent to the bottom of the river by placing stone masonry or concrete on the roof. Compressed air piped to the sunken box kept the river out, helped keep water pressures from crushing the box and made it possible for men to go down inside to dig out the river bed.

As digging progressed the box would sink slowly, steadily, deeper and deeper, until it hit a firm footing. Then excavation stopped and the interior of the box was filled with concrete. This served as the foundation for the bridge piers. Fortunately, engineers have since developed better and much safer construction methods for these difficult foundation conditions.

An inspector does a final check of the reinforcing steel inside the footing rebar cage at pier 9.

place cofferdams at the other river piers, the steel bracing frame of pier 9 was prefabricated on a barge, floated to the location and lowered to the river bottom.

The cofferdam frame utilized tubular sections for the vertical members so that additional tie-down anchors could be drilled through the tubes into the bedrock if cofferdam stability had been questionable. The stability of the cofferdam was additionally improved by placing 480 tons of riprap on the outside of the cofferdam around the north edge, which also prevented further scour from taking place.

Interestingly, contractor Meritt-Chapman, who had constructed the 1911 bridge 60 feet to the north, had also experienced his biggest construction problems at pier 9, with river scour to a depth of 25 feet having been reported during construction.

Scour was carefully monitored at each cofferdam throughout the construction. Scour on piers 3, 4 and 5 was slight — only a few feet of scour. At piers 7 and 8, however, scour measured up to 20 feet and was most evident on the north (upstream) end of the cofferdams. The scour at pier 7 was believed to be a major contributor to problems encountered during the placements of the reinforced footing for that pier.

On April 13, 1991, when about 400 of the 2,000 yards of the foundation concrete had been placed at pier 7, water broke through along the edges of the north end of the cofferdam. Placement was stopped and several weeks were lost to remove the water-damaged concrete. A perimeter drain was installed to stabilize flow and the remainder of the footing was placed without further incident. The remaining two piers, 8 and 9, had the same type of perimeter drain system installed as a precautionary measure, but there were no further problems with leaks.

Temporary fender system

Before men would be sent 30 feet below water level to work in the cofferdams adjacent to the navigational channel at piers 8 and 9, a temporary fender system was constructed on the channel side of these cofferdams. This was to protect the workers in the unlikely event a work barge or some other vessel collided into the cofferdam. The fenders were subcontracted to Blakeslee Arpaia Chapman of Branford, Connecticut.

There were other safety measures taken, too. For example, a rescue boat, with its engine running and at least one CPR-trained person on board, was stationed at the bridge whenever work was going on over the water. Handrails were installed wherever feasible and life jackets were worn by those working on or over the water. To the contractor's credit, there were no major accidents during the construction process.

The construction of the west fender adjacent to pier 8, using timber piles, was carried out in June of 1990 essentially without problems, and the work moved to the east fender.

When pile-driving began on the east temporary fender, it was discovered that the depth of overlying river bottom material could not support the piles — let alone resist barge impacts — particularly at the south end. An alternate plan was needed quickly because pier 9 cofferdam construction could not begin until the east fender was in place.

The solution was to make use of the caisson technology that was to be used for the foundation at pier 9. To further expedite the work and because the caissons were in the open river without cofferdam shells around them, Blakeslee proposed a mechanical stiffening fin arrangement on the pile tips in lieu of grouting to stiffen the interaction with the bedrock. Work on the east fender proceeded.

Timber piles being installed at the west temporary fender by Blakeslee Arpaia Chapman. Fenders were installed adjacent to the navigation channels to protect men and equipment working on and below water level.

While there is evidence today of some minor collision damage — most likely from smaller vessels or construction barges — fortunately neither fender was put to a critical test during the cofferdam installation. Both fenders will be removed as part of the contract to demolish the existing bridge, to provide the 200-foot navigation channel width the Coast Guard requires.

Labor crews clean all laitance from the top of the tremie seal before the rebar crew from Advance Steel moves in to tie the footing rebar cage. In this photo, the piles have all been cut to provide a level 12 inches of penetration into the footing. The bottom layer of reinforcing steel will lie directly on top of these pile tops.

Pier bases and granite fascia

At each pier, once the foundation concrete was placed, work could begin on the next level — the pier base. The foundation had the same shape as the cofferdam itself — in fact, the cofferdam was used as the form for the foundation. Two rows of 12-inch studs welded onto the sides of the inside walls of the cofferdams ensured that the cofferdam and the foundation concrete would act together. The pier base, however, would extend above the normal river bottom levels and was shaped with rounded ends at the upstream and downstream directions to impede river flow as little as possible and minimize scour action.

The construction of the pier bases would be in one of two lifts — a "lift" being the height of the forms (or panels) into which concrete is poured. The shallower piers would have just a single lift; the deeper piers (pier 5 to pier 9) would have two lifts. All would rise to a point about five feet below water level. There, four courses of granite stone — each $2\frac{1}{2}$ feet high — would add another 10 feet of height and bring the pier base to an elevation of five feet above mean sea level. The normal tidal action, therefore, would always be in the range where the granite stone was placed. The granite looks aesthetically superior to bare concrete and structurally it provides superior corrosion and abrasion resistance for tidal flow year-round. It is particularly resistant to the ravages of ice in winter.

Pier shafts

Up to this height in the bridge construction process, the northbound and southbound lanes have shared a common substructure; above the granite fascia, the columns and the surface deck are completely independent.

Each of the twin rectangular columns measures 8 feet by 36 feet. They were placed in up to four lifts with heights up to 20 feet with construction usually taking 8 to 10 days per lift. PPO used three sets of column forms and one cap form for the top section. These forms were raised and lowered by barge-mounted 200-ton capacity cranes and moved generally from the west piers to the east piers over a period from September, 1990, to February, 1992.

All of the concrete — some 54,000 cubic yards for the substructure alone — was delivered by truck from the Tilcon Connecticut ready-mix plant just off the Boston Post Road in Old Saybrook, about a mile from the site. A backup plant at Portland was put on standby at each major placement in the event of a breakdown, but its services were never needed.

The contractor had initially considered the possibility of setting up its own concrete plant to supply all of the concrete for the substructure as well as for the 32,300 cubic yards needed for the precast segments. However, the economic analysis of this option versus utilizing an existing plant only a mile from the job site clearly favored using Tilcon, particularly after the car barges were extended out onto the river and Tilcon transit mixer trucks could drive right to the piers. The choice was a good one, and very few problems with concrete supply or quality occurred — a credit to Tom Small of Tilcon who was also responsible for the concrete mix designs for each different element of the structure.

One major change that Small proposed was the use of liquid nitrogen to cool the concrete on its way to the job site, rather than replacing the water in the concrete mix with ice for the tremie placement as the contract specified. (Ice is often used to keep the heat from hydration of the concrete as low as possible in the massive placements in an attempt to control potential thermal cracking of the concrete.)

Small believed that the ice might not melt fully prior to placement and would lead to problems with workability of the mix, inconsistency and possible strength loss. His proposal to use the liquid nitrogen was accepted, and the criteria for cooling was modified such that the temperature of the concrete at delivery was always within 10 degrees of river temperature. This was important because if the exterior cools too fast, temperature differentials during the hardening process can lead to cracking.

The nitrogen plant was set up on the site of the old ConnDOT Maintenance Facility just off of Ferry Road by Airco Industrial Gases and was in operation from August,

Above the shared footing and pier base, two separate columns rise to support the bridge superstructure. In this photo, the forms have been raised on both columns to the second lift. Rebar protruding above the top of the form will extend into the next lift of the pier column.

Billowing clouds of nitrogen signal the loading of another truck headed to the bridge. The liquid nitrogen had to be added in short bursts to keep the cold temperatures from cracking the steel shell of the mixer trucks.

TO ENSURE uniform and reliable properties of concrete delivered to any project, mix design parameters are specified in the contract to ensure that the designer's assumptions, particularly about strength, are met. Tests are conducted at the site to measure temperature, air content and slump (which is an early indicator of excess water in the mix). The mix itself is very tightly controlled at the plant, which is semiautomatic and largely computer controlled so that mix problems are very rare.

The accepted standard for concrete strength is known as the 28-day compressive strength test. Though concrete actually continues to gain strength for many years, minimum strengths are specified at 28 days. The test involves measuring the force it takes to crush a six-inch-diameter by 12-inch-high test cylinder of concrete. Usually the average of two tests is taken as its strength.

The curing process of concrete is interesting and frequently misunderstood.

Concrete is a mixture of Portland cement, sand, crushed rock and water. Strength is achieved when the Portland cement reacts with water in the mix; maximum strengths don't occur until all of the water has been consumed in this reaction process. If concrete is poured on dry and windy days, some water is lost to the atmosphere which results in shrinkage cracks and loss of strength. Fresh concrete is best covered with saturated curing blankets in the first three to five days after placement and kept wet to avoid this problem. Surprising to those unfamiliar with concrete, is that the most ideal condition for curing concrete (after the initial set of the concrete has taken place), is under water.

Element	Type of concrete	Required 28-day strength
Tremie seals	Class 'A' (underwater)	3,000 psi
Foundations	Class 'A'	4,000 psi
Piers	Class 40	4,000 psi
Precast segments	Class 55	5,500 psi
Parapets	Class 'F'	4,000 psi

A six-inch-diameter concrete test cylinder is crushed in an axial compression testing instrument to determine the compressive strength of the concrete. Of importance to the construction schedule is not the strength at 28 days, but the strength at one or two days, which will dictate when the forms can be stripped. In normal temperatures, a mix designed to have a minimum of 4,000 psi at 28 days might be expected to have around 1,500 to 2,000 psi at 24 hours, which is usually sufficient for form removal.

1990, until September, 1991, when it was dismantled and removed.

The launching gantry

As foundation work on the bridge proceeded, other elements of the construction process were moving ahead, too. Detailed shop drawings and working drawings were being prepared by a host of subcontractors, suppliers and consultants with various areas of expertise.

Perhaps the most important of these was being carried out by the Denver firm of Finley McNary Engineers. They were given the task of preparing all the shop drawings for the precast work and for developing a step-by-step manual for the erection of the segments, together with a check of structure stresses at each stage.

Under a subcontract agreement to Finley McNary, the design for the gantry itself would be undertaken by the French design firm of Freysinnet — probably the most experienced designer of launching gantries in the world.

Eugene Freysinnet, the founder of the firm, is credited with popularizing prestressed concrete construction in the late 1940s. Actually, prestressed concrete concepts were formulated as early as 1886, but their application had not been successful because of the unavailability of suitable steel. In the mid-1920s, Freysinnet solved the problem by using newly developed high-strength steel, which made prestressing a viable means of construction.

Interesting, too, is the fact that Charles Rejcha, who was responsible for Parsons Brinckerhoff's design concept for the Baldwin Bridge, worked in Freysinnet's Paris office in

J. (Skip) Conti and William Sasser, both of ConnDOT's District II Construction, were the department officials most involved in the new bridge construction. Sasser was assigned full time to the project; Conti, as supervising engineer, was responsible for coordinating the bridge construction and the two approaches.

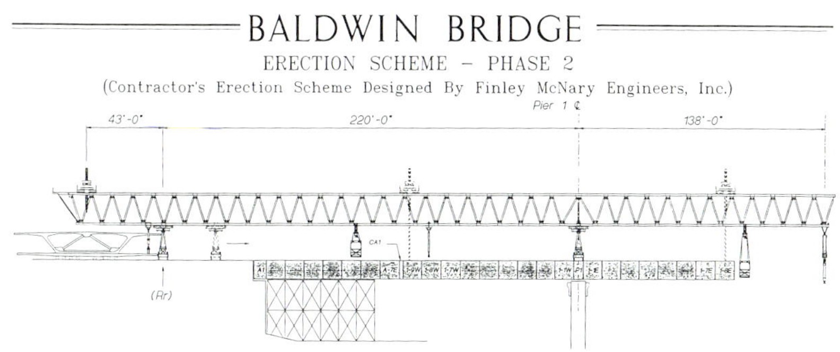

A drawing of the launching gantry shows its size relative to the bridge.

The launching gantry was assembled on the west approach embankment in the spring of 1991. When sections started arriving at the site, many local residents thought the brightly painted red and white pieces were parts of the new bridge itself.

1947. Ten years later, he immigrated to the U.S. where he worked for the Freysinnet Company of New York, eventually as president of its engineering subsidiary before joining Parsons Brinckerhoff. He has since retired, but his interest in the project remained. He was a frequent visitor to the job site just to "see how things were going," often stopping by while enjoying his favorite pastime of boating on Long Island Sound.

PPO and Finley McNary closely monitored the Freysinnet design process and by August 1990, the design was complete and drawings were sent to prospective fabricators for competitive bids. The successful bidder, which interestingly was not the lowest bidder, was the joint venture of J.O. Dougal and Canron from Ontario, Canada.

Most often a contractor will solicit competitive bids and will usually take the low bid. In this case, PPO was acutely aware that the launch gantry operation was key to timely completion of the project and required each bidder to commit to a schedule. The bids were evaluated by PCL engineer Alan Brayley, who made the final recommendation.

Brayley explains, "It was vitally important that we choose an experienced fabricator with the engineering and production capabilities to deliver the gantry on time. We considered the three lowest bids carefully and concluded that the premium to be paid by selecting the second-lowest bid from Canron/Dougall would be more than offset by the quality of work and their guarantee of on-time delivery."

Fabrication of gantry components was carried out through the winter of 1990-1991 and gantry pieces started to arrive on site early in April, 1991. Many local residents thought the brightly painted, red and white pieces were actually parts of the bridge. The gantry, which was 439 feet long and weighed about 600 tons, was assembled on the west approach over a 10-week period. By the end of June it was almost ready to begin its tour of duty — soon it would be time to "hang segments."

The Parsons Brinckerhoff/L-C Associates inspection team involved up to 22 engineers, inspectors and administrators at the peak of the project. Five of the team had worked together before on the Sunshine Skyway Bridge in Florida.

Casting yard operations got underway late fall, 1990. There would be 488 precast segments cast over a 19-month period without a single reject. Here a previously cast segment is being positioned for use as the form face for the next segment to be cast — ensuring a perfect fit when the two are joined together again on the bridge.

CHAPTER 6 Precasting Operations

WHAT DIFFERENTIATES the Baldwin Bridge from other bridges in the state, apart from the sheer physical size of the bridge, is not in the substructure, but in the deck construction.

In a more conventional structure, the foundations and piers are constructed and then the deck is built on top. In a precast bridge, however, the deck sections or segments as they are normally referred to, are fabricated or cast in another location while the substructure is being built. Then the deck segments are erected directly on top.

Clearly, there can be substantial savings over conventional methods if a system can be developed for high-speed erection of segments. Concrete segmental bridges have rapidly gained popularity in the U.S. and around the world as they lend themselves to very rapid assembly — particularly when erected with a launching gantry. While work on the foundation had begun early in the spring of 1990, casting-yard operations began just a few months later, in July, 1990.

Finding a location for the casting yard proved to be quite a bit more difficult than expected. At bid time, the contractor planned to use the Island Cove Marina in Old Saybrook, just $\frac{1}{4}$ mile upstream from the job site. However, the zoning permit request to change from commercial to industrial use was denied, probably because of possible disruption to the adjoining marinas and the fact that casting yards are not particularly pretty.

The search widened considerably. PPO looked at possible sites for the yard from the New Jersey coast to Quonset Point in Rhode Island, which was serving as the casting yard for the precast segments of the Jamestown Bridge.

In fact, the location of the casting yard and the method of transporting the segments to the job site was a key factor in the contractor's decision to use a launching gantry. Proximity was doubly important because PPO knew they'd have to truck the segments from the yard to the bridge.

Recognizing the contractor's difficulties and in an effort to support the alternate erection scheme and the accompanying early completion date, ConnDOT offered use of a five-acre portion of land adjacent to Ferry Road in Old Saybrook. The yard had been the site of their old maintenance depot and it was scheduled for demolition under the west approach contract.

A rebar cage, complete with all the internal post-tensioning ducts, precast struts and edge beams, is lifted out of the rebar jig and prepared for placement in an empty form as part of the assembly line process.

Discussions were underway when another option developed and zoning approval was obtained to set up the yard on a 15-acre wooded lot between the Boston Post Road and Amtrak, less than a mile from the job site. However, the clock was running and it was into July, 1990, before equipment was able to move in to start clearing and grubbing.

The casting yard, which can be likened to a factory assembly line for precast concrete segments, was given priority over the next four months. A 120-foot-high tower crane on rails was assembled. Three casting machines were installed. Six rebar jigs to pre-assemble the rebar and post-tensioning ducts were fabricated. A stretch fabric building was assembled for storage of material which required protection from the elements.

On Wednesday, November 21, 1990, just one day before Thanksgiving, in freezing weather with snow flurries in the air, the first of 488 precast segments was cast. Over the course of the next 19 months, 487 more segments would be cast without a single rejection. Much of the credit for this remarkable achievement must go to two men; Bob Novynka and Gary Fenton, who work for PCL.

Bob Novynka remembers those early days. "We got off to a slow start in the casting yard," he says. "Since this was the first segmental bridge in the state, there were only a handful of people who really knew how to do this type of work. Once everyone gained some experience, including the inspection group, we began to work as a team. That made things go a lot better. It took a while, but in a few months, we'd come through our learning curve and got to the production rates we'd planned. Some doubted our ability at first — but we proved ourselves in the final count."

The method of casting these segments is not new and has evolved over the past three decades from its origins in Europe — most notably in France. The particular method used at the bridge is referred to as "short line" casting. In this process, each segment is cast between a steel bulkhead and the segment cast previously. The starter segments on each line — usually the pier segment — are cast between two steel bulkheads.

The casting yard had three casting beds. One bed was used to cast pier segments (the ones that sit directly on top of piers) and abutment segments (the four segments at each end of the bridge). There were a total of 24 of these two segment types.

A second bed was used to cast the thick-walled type "B" segments which butted up to the pier segments and extended two or three segments away from piers in each direction. There were 76 "B" segments.

The third casting bed was used to cast the type "A" thin-wall segments (a total of 388) which were used everywhere else. To speed things along, the second form was converted to cast type "A" segments in October, 1991, after all the thick-walled type "B" segments were cast.

Type "A"segments could be turned out on a daily basis. Type "B" segments could normally be cast every other day, and pier segments took an average of 7 to 10 days to cast. Extra time was required because they had a different internal shape, were more complex and required the rebar and post-tensioning ducts to be tied in place.

The concrete used in the precast segments was a higher strength than that required for the substructure. A concrete strength of 5,500 psi at 28 days was specified versus 4,000 psi

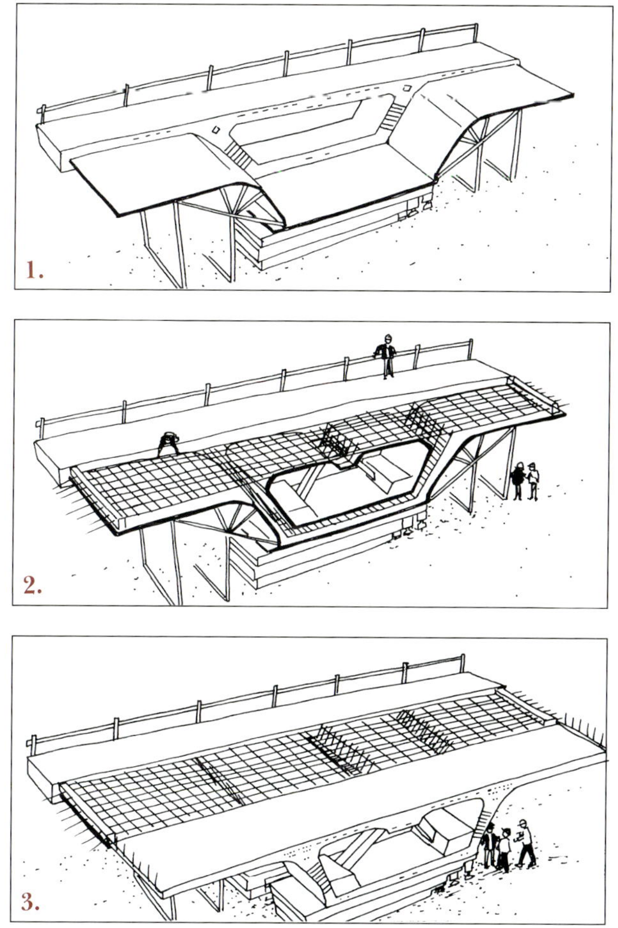

This sequence shows how segments were made in the casting yard. (1) The form and the steel bulkhead mold behind it are cleaned and prepared. (2) A rebar/post-tensioning cage is lowered into the empty form. (3) The previously cast segment (match cast segment) is moved into position. Concrete is now poured into the form between the match cast and the steel bulkhead, creating the new segment.

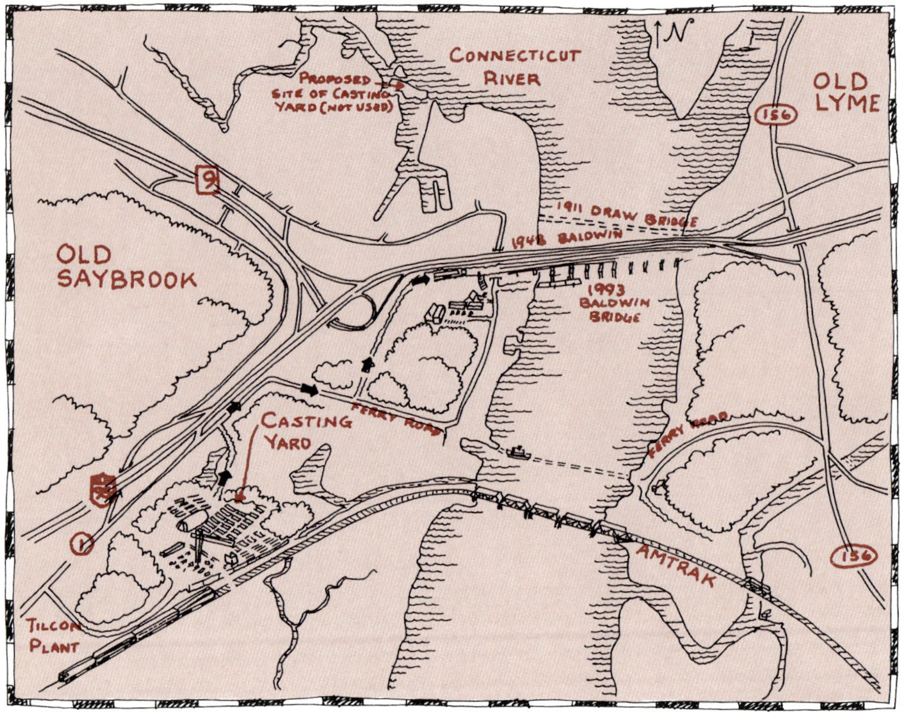

General layout of the area showing the railroad and the old ferry route, the location of the 1911 bridge, the casting yard and the segment delivery route. Also shown at the top of the map is the marina originally planned for the casting yard before zoning problems were encountered. The high-altitude photograph on page 6 can also be used to view these locations.

for the pier columns and foundations. The designer required a higher strength for maximum efficiency of the section and minimum weight. This was obtained by using water-reducing additives and increasing the cement content per cubic yard.

Up to a point, concrete strength increases with increased cement weight per cubic yard. Strength can then be carried higher by chemical additives such as superplasticizers, which allow water content to be reduced while still keeping the fresh concrete in a fluid state.

The engineers at the bridge site kept a close eye on the casting yard operation since they were the ones who'd be responsible for assembling the pieces. Any mistakes that went undiscovered would be very costly. The key was to get it right the first time, and then double-check just to be sure.

The segment casting began four months later than the planned start and went through a slightly longer than anticipated learning curve to get up to full production. After that, however, casting proceeded almost without problems at about 30 segments per month working through two winters until the last segment was cast on June 26, 1992 — two weeks ahead of the original scheduled completion date for casting.

Steve Gage of the ConnDOT Rocky Hill lab, in charge of the casting yard inspection team, talks about ConnDOT's first experience with precast segmental construction. "It was a new process for us," he says, "but very educational. What really impressed me was the exemplary cooperation and open communication between everyone involved. Compared to some other precast yards I've seen, the quality of the work was exceptional. If I had to do it again, I wouldn't change a thing."

John O'Brien, who was directly in charge of the casting yard operations for ConnDOT, reaffirmed Steve's perception and added that the contractor's approach to safety and safety record were noteworthy.

The contractor was required to prepare detailed "shop" drawings. The purpose of these drawings was to show, in greater detail than the contract drawings, all the various design details for each portion of the bridge to make sure that everything would fit together.

Essentially PPO had to build it on paper and then build it again in the field. The contractor was also required to prepare technical submissions analyzing the effects of proposed construction loads on the partially completed bridge. Most of this work for the superstructure was performed by consultant Finley McNary.

The review of these shop drawings and technical submissions was carried out in Parsons Brinckerhoff's Tampa office, which was originally responsible for the design of the segmental superstructure. John Allen led a team of engineers in checking almost 2,000 construction drawings, the majority of which were reviewed in the first six months of 1991.

The journey from casting yard to job site

To get cast segments from the casting yard to the bridge, the contractor was granted a permit to transport the segments on local roads, the Boston Post Road and then onto Ferry Road. The permit for each trip was to cost $23. Oddly enough, the initial agreement with the permit section of ConnDOT required payment to be made each time a trip was made. Later, when up to eight segments would be

LABOR AND LABOR RELATIONS

WHEN THE casting yard was at full production, the contractor's work force totalled about 75 employees, including ironworkers, carpenters, concrete masons, laborers and operating engineers, supervised by a handful of superintendents and engineers. Work force at the job site in the fall of 1991 when the erection had begun and the substructure was still ongoing was about 90 people. At its peak, around 165 men and women were employed by the contractor and subcontractors.

Behind the scenes there had been quite some discussion between PPO and the local trade unions over which trade would erect the segments. The carpenters and the ironworkers both claimed it was theirs by right. The Connecticut Labor Department was consulted and the contractor was given the freedom, since no precedent existed in the state for this type of work, to allocate work tasks.

In the end, the work was split. The carpenters would be responsible for handling and gluing the segments and for all gantry moves, while the ironworkers would perform all the post-tensioning and rebar installation. Both crews would have to work together. For the most part, it worked quite well.

The tower crane operator in the casting yard had a birds-eye view of the precast segments awaiting their trip to the bridge in the distance. The smaller 300-ton travel crane picked segments out of storage and loaded them onto the segment hauler to be trucked to the bridge site.

moved in a 24-hour period, a quarterly payment schedule was established. The permit required that the hauler would always have three escort vehicles for the trip — two at the rear and one in front.

The original plan called for exiting the yard onto Roamtree Drive, following Boston Post Road to Ferry Road, and then up the haul road to the bridge. But a new entrance to I-95 from the Boston Post Road had just been completed and heavy traffic, especially in the summer months on Friday afternoons and Saturday mornings, might disrupt segment hauling — and vice versa.

Old Saybrook police recommended changing the route to go through the Gladeview Healthcare Center. With the concurrence of everyone involved, the new route was used with obvious benefit to all concerned.

Getting the haul road built required more negotiation. After leaving Ferry Road, the last leg of the trip was along the side of the new I-95 on-ramp from Route 9 which was being constructed by the Middlesex Corporation as part of the west approach. The temporary haul road would have to be built through Middlesex's project limits and PPO had to strike a deal with its neighbor. ConnDOT participated in early negotiations and then left the contractors to work out details, which they did.

On the afternoon of June 18, 1991, almost seven months after casting began, final preparations were made for transporting the first segment to the bridge site. Segment No. 108 was placed on the special 12-axle, 96-wheel hydraulic trailer late in the afternoon and last-minute checks were carried out. Everything seemed in order. At dawn the segment would make the milestone trip.

At 5:45 a.m. on June 19, the 475 HP tractor and trailer unit, 110 feet long with a combined gross weight of approximately 270 tons, slowly eased out of the yard, followed its course through the deserted streets of Old Saybrook, climbed the haul road and backed up to the rear of the gantry which was positioned on the west approach. Except for those directly involved, and a handful of engineers who got up early to witness the move, the event had gone almost unnoticed.

Chosen for the first trip because it was one of the heaviest segments, No. 108 would be used for the load test. The pier segment for pier 1 would be the next segment brought "to the hill" and would be the first segment to be erected.

Before beginning segment erection, the engineers, who are cautious people by nature, devised a series of tests to be sure the launching gantry would perform to specifications — if problems were going to develop, better to discover them 10 feet above the west approach than 105 feet above the Connecticut River.

The test involved having the gantry move the heaviest segment to a position that created the most severe load condition possible. The test actually involved a situation more critical than any that would actually be called upon. The test was successfully run, then repeated with an additional 10 percent superimposed load. Deflections, recorded by two teams of surveyors, came within an inch of the theoretical cantilever deflection of 24 inches. All components were rechecked for signs of overstress, and none was found. Final field modifications and assembly were completed, and the gantry was launched to pier 1 on July 10, 1991, with considerable local public interest and curiosity.

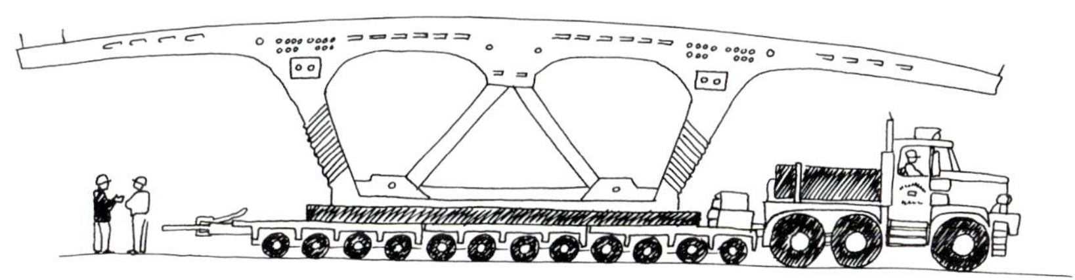

A 140-ton segment, pulled by a 475-horsepower tractor, makes its way from the casting yard to the bridge via Gladeview Healthcare Center. The trailer used to transport the segments is not unique; units like this are commercially available — at a price. All 12 axles turn independently and in synchrony each time it goes around corners, greatly increasing its maneuverability.

The first pier segment is placed on top of pier 1 column bearings on July 13, 1991. (Insert) Robert "Kimo" Avery, PPO's survey party chief responsible for, among other things, ensuring that the segment would fit on top of the pier — and it did.

CHAPTER 7 Erecting the Segments

TWO DAYS after final testing of the gantry, on the afternoon of July 12, PPO began erection of the first pier segment. With the segment picked just a few feet off the ground, however, a hydraulic line failed, spewing hydraulic fluid over the east face of the segment.

Fortunately, the gantry's lifting hydraulics had been designed with many safety features including check valves which would automatically close upon sudden loss of pressure. The faulty line was replaced and a thorough recheck performed. The segment was carefully cleaned with high-pressure water-blasting equipment and solvents — any residual hydraulic fluid might affect the bond of the epoxy joining this segment with the next.

Erection of the segment was postponed to the following day — Saturday, July 13. PPO felt the minor bad luck they'd had with this segment was past and were less concerned about superstition than about how the segment would fit on the pier. To be sure, placing a 140-ton piece of concrete 60 feet in the air and assuring the plus or minus $\frac{1}{16}$ inch tolerance was on everyone's mind — none more so than PPO survey team party chief Robert "Kimo" Avery of Windham, Connecticut, and his brother, Barry. They would be the ones to answer if things didn't quite fit.

Asked what he remembers of that day and if he was concerned, Kimo replied with quiet, unassuming confidence, "Actually we weren't really too worried. We'd carefully surveyed the pier cap, the location of the earthquake restrainer and pads for the disc bearings. Everything seemed to match perfectly with the survey that Ian White and his casting yard survey team had taken earlier on the pier segment we were placing. There was a lot going on that day and we were pretty busy. We were actually more concerned with the ability of the gantry to be able to locate and adjust the segment to its resting point — which proved to be the biggest challenge. Once it was in position, I remember someone saying, 'O.k. — one down, 487 to go!'"

By the end of the day, the segment was sitting firmly on the pier bearings and within $\frac{1}{4}$ inch of final position. Final alignment would be done on Monday and the second segment could then be brought out and attached to the land side of the pier segment.

Segment erection had begun. A handful of local observers had noticed the spectacle of the segment slowly moving out high over Ferry Road and had stopped to watch. Flagmen were used to stop traffic as the segment crossed the road. This fascination of the public over the bridge "erector

DISC BEARINGS

THE DISC bearings sit on top of concrete pedestals typically about 15 inches above the top of the pier cap. The bearings are responsible for transferring all of the superstructure loads to the pier caps and into the substructure. They must also allow the bridge to "breathe" freely under temperature and long-term deflections which, if restrained, would cause unnecessary and detrimental stresses to develop. This is done by equipping some of the bearings with teflon-coated sliding plates so that vertical load is carried but horizontal loads cannot develop. To be sure this freedom only occurs where the designer intends it, other piers are fully fixed horizontally while some have freedom in one direction only.

Traffic on Old Saybrook's Ferry Road is stopped for a few minutes as a segment is transported by the gantry crane over the roadway during the erection of the first span. A few local residents stop to admire the spectacle.

set" erection would continue for the next $14\frac{1}{2}$ months. For some it turned into a hobby and they became frequent visitors to the site, establishing a relationship with the site personnel. Retired engineer Jack Greenwood, of Essex Meadows, was one of the project's most faithful visitors.

Jack recalls his first fascination with the project. "I was just amazed at the mass of equipment on the river for the foundation work. As the pier columns started to rise out of the water and the segments started to be delivered, the whole project became engrossing. Nobody from the area had ever seen anything quite like it, and nobody could quite believe the speed at which construction moved ahead — particularly the second crossing of the launching gantry. It was a marvel to watch. We're all looking forward to crossing the new bridge."

After the pier segment is placed and clamped vertically to the pier, the next segment is attached to one side of the pier segment. Because each segment was cast using the face of the adjacent one as a mold, the segments match perfectly. The face of each segment has alignment keys and shear keys. The alignment keys in the top and bottom deck sections help during segment erection to bring the two segments into the same orientation they had while being cast. Later, the top deck alignment keys also serve to distribute traffic loads across the joint. The shear keys in the webs of the segments transmit shear or vertical loads through the joint.

Each segment is initially clamped to the preceding segment with eight $1\frac{1}{4}$" diameter high-strength post-tensioning bars, each stressed by hydraulic center-hole jacks to 50 tons clamping force. The bars are installed loosely in the segment to be erected while it is still on the trailer, usually at the same time the lifting frame from the overhead crane is attached. When the segment is brought out and while the epoxy adhesive is being mixed and applied, the threaded post-tensioning bars are coupled to the eight bars in the previously placed segment.

The launching gantry's overhead crane moves the segment backwards to close the joint and the eight hydraulic jacks are attached to clamp the segments together. Anchor plates and nuts on the end of the threaded bars lock in the clamping pressure of about 40 psi across the approximately 140 square feet of the joint.

With two segments now in position, but with the second hanging in a seemingly unsupported position, visitors frequently asked, Why doesn't it fall over?" The obvious imbalance at this point in construction is carried by two hydraulic jacks sitting on temporary supports which are fastened to the pier column cap.

Next comes placement of the segment's symmetrical partner on the other side of the pier. With the pair erected and balance restored, additional post-tensioning forces are applied using much larger hydraulic jacks and high-strength wires commonly referred to as strands.

Segment erection continued in similar fashion until all the segments that would rest on a particular pier were in place. Called a fully balanced cantilever, this stage resembles a giant see-saw with the pier top as the pivot point. Great care was taken to maintain its balance and stability during construction. The legs of the launching gantry carried some of the imbalance to its supports; a 125-ton concrete counterweight did the rest. This 125-ton concrete block was nicknamed Big Bertha and it was colorfully decorated by the

THE EPOXY ADHESIVE

PRIOR to joining the segments, about 10 to 12 gallons of epoxy are mixed and applied to the face of one of the segments. Sikadur 31 SBA, a specially formulated two-part epoxy adhesive for bridge construction was used.

Carpenters then mix the resin and hardener which is applied to the joint by gloved hand. Care is taken not to get excess glue into the post-tensioning ducts which could result in blockage when tendons are later pulled through.

The glue initially acts as a lubricant to allow the segments to slide easily into place. Later it acts as a moisture barrier to prevent water entering the joints and corroding the post-tensioning tendons running through the joints. It also serves to distribute compression from the post-tensioning uniformly through the joint. Actually, it has considerable tensile strength. The design, however, does not rely on any glue strength. The structural integrity comes solely from the clamping action of the post-tensioning tendons.

The glue in the joints is compressed by the effects of the post-tensioning to a nominal $\frac{1}{32}$ of an inch. When work progressed over Ferry Road, traffic was redirected so the excess globs of epoxy being squeezed out wouldn't fall on passing vehicles. As work extended over the river, a cloth "diaper" was draped across the full length of the joint to catch the glue and prevent it falling into the river or onto passing boats.

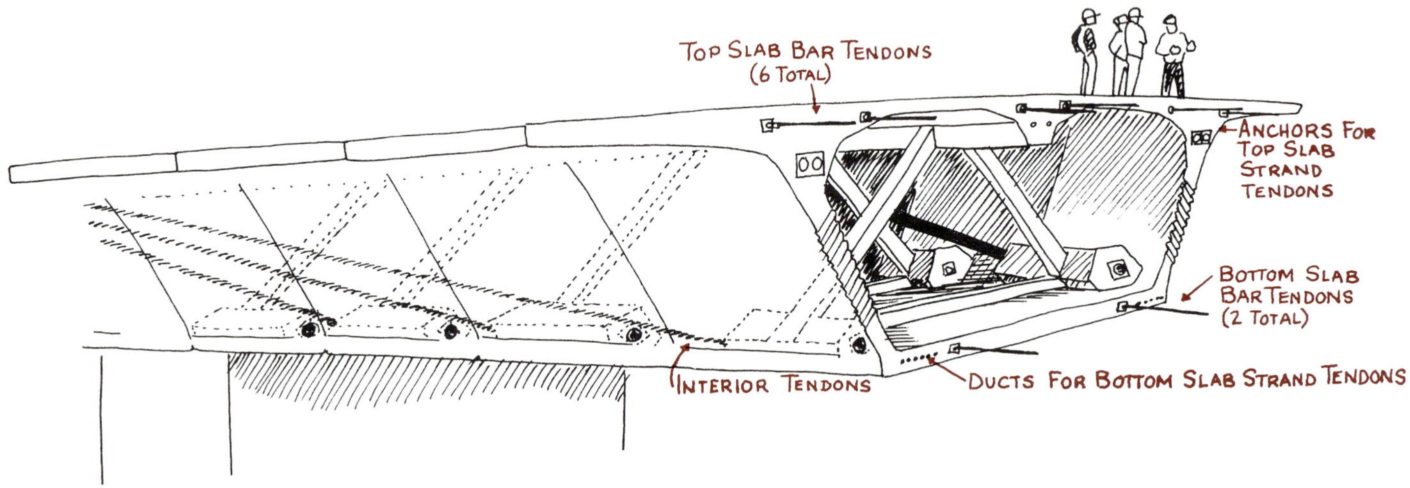

Tendons are principal means of holding the bridge together, especially during erection. Longitudinal tendons run in the same direction as the bridge. Some connect each segment to its match cast segment; others connect it with the pier segment and its mate on the other side of the pier. Transverse tendons (running across the width of the bridge) are not shown for clarity. The transverse tendons are post-tensioned in the casting yard before the segment is transported to the bridge; the longitudinal tendons are post-tensioned during erection.

survey crew during slack time. It is moved several times in each cantilever cycle by the overhead crane.

Closure pours

When work progressed to the point where all segments had been placed between two piers, there remained a small space which needed to be filled. These "in-place closure pours" were made with concrete brought by truck from the Tilcon plant along the same route taken by the segment hauler.

The two cantilevers are aligned, both vertically and horizontally, to be sure the joint is perfect. Panel forms are then placed to the same identical cross-section as the segments, a reinforcement cage is tied in place and concrete placed in the joint.

By the following day, sufficient strength (2,600 psi minimum) has usually developed so that post-tensioning of the tendons through the bottom slab across the joint can begin. These tendons vary in length from the shortest, which goes between the segments on either side of the cast-in-place joint, to the longest tendons between adjacent piers. These tendons serve to resist positive moments (SAG moments) which will result from traffic.

When all the tendons across the closure joint are tensioned, the erection crews prepare to launch the gantry to the next pier and start the cycle all over again.

Behind the erection crew comes the grouting crew. The grouting crew's job is to grout and cap all the post-tensioning tendons after they are post-tensioned. The grout is pumped from a mixer into the ducts through which the

STRANDS AND TENDONS

EACH STRAND is made up of seven wires, six of them coiled around a central core wire. This material has a very high tensile strength guaranteed not to fail at stresses up to 270,000 psi or 135 tons per square inch. All of the post-tensioning strand and hardware was supplied by VSL from Los Gatos, California.

A tendon is made up of a number of seven-wire strands — up to 27 strands per tendon. The tendons are made up in advance by the ironworkers and laid out on the previously erected deck. The leading edge is welded together and an eyelet welded on to enable the strand bundle to be pulled through the post-tensioning ducts by a "tugger" on the leading end of the deck cantilever. They are then anchored with gripping wedges on threaded nuts as explained on page 43.

Typical tendon forces are in the range of 450 tons each. There are normally a minimum of two and up to five of these tendons installed at each joint, dependent on the location in the span. As the spans are put together, all of these tendons run through the top slab of the segments to resist negative moments (HOG moments) from the structure's own self-weight.

Ironworkers on the stressing platform tension bar tendons as the segments are erected. They later stress the strand tendons, clamping the segment to its corresponding partner on the other side of the pier segment. The pipe on the right is used to pull tendons across the open gap into the post-tensioning ducts during installation.

L-C Associates inspector Jim Bergeron checks tendon elongation during a strand-tensioning operation. Field-measured elongation must agree within 5 percent of theoretical elongation as a check on the force applied by the hydraulic jack which is measured by gage pressure.

tendons pass. These are only about half filled with strands or bars so there is room for a substantial amount of grout. The grout acts as corrosion protection for these critical strength members of the bridge.

Placing the abutment segment

An exception to the typical cantilever erection occurs with the abutment segments on each end of the bridge. These segments are erected on temporary support piers, called falsework because there are no balancing segments on the other side of the pier as is normally the case.

When each segment is lowered onto the falsework, it is actually sitting on three 100-ton hydraulic jacks, each interconnected to each other and to jacks underneath other segments on the falsework. This compensates for any settlement that might take place during installation. After all 11 of the segments in the half span of the abutment are set on the falsework and aligned correctly, they are tensioned together by the same eight post-tensioning bars used on

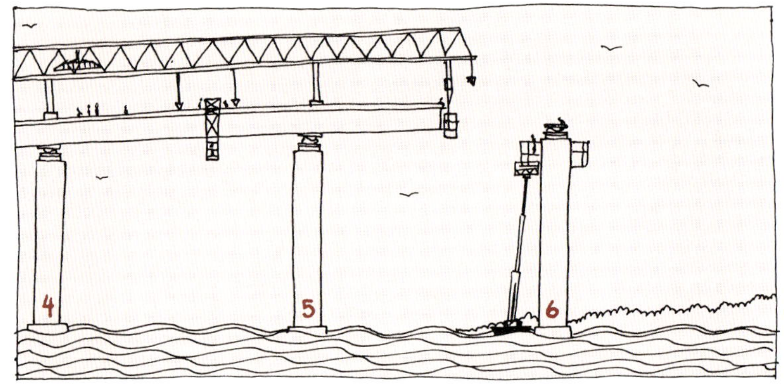

1

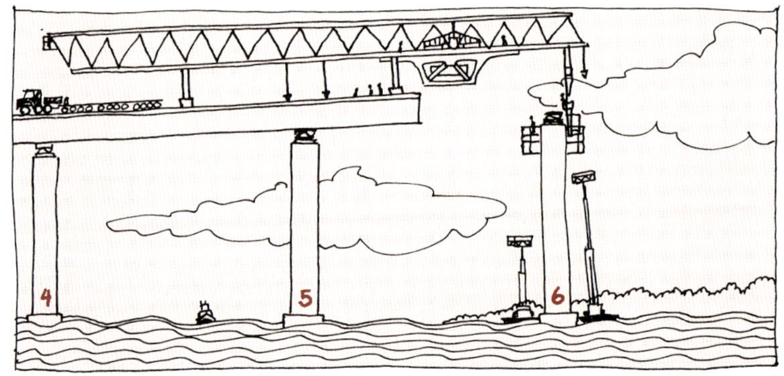

2

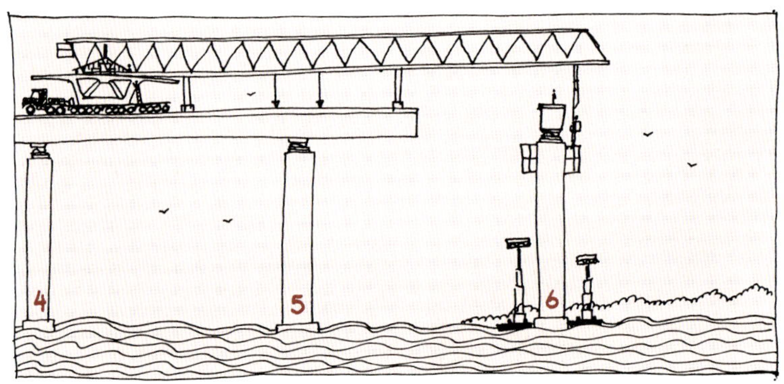

3

(1) A closure pour is placed midway between piers 4 and 5. Preparation is ongoing at pier 6 setting the bearings and doing final survey work to be sure everything will fit.

(2) The launching gantry moves to pier 6 and a temporary support is attached to the front face of the pier. The pier segment is unloaded and the launching gantry's crane carries it to pier 6 where it is placed on the bearings.

(3) The pier segment is clamped by vertical post-tensioning bars to hold it securely in place. The next segment is delivered to the rear end of the gantry and hooked up to the crane.

other segments; then a cast-in-place closure joint is made.

The westbound structure was built first, taking ten months from mid-July 1991 to mid-May 1992. Erection continued throughout the winter but not without problems.

Wintertime problems with the epoxy

As the cold weather moved in by mid-December of 1991, potential problems with the epoxy adhesive surfaced. With normal temperatures, the epoxy gained its strength quite

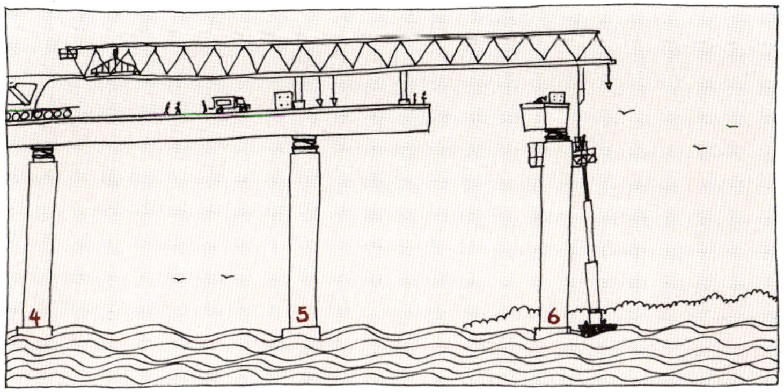

4

(4) The second segment is erected and clamped to the pier segment with eight post-tensioning bars, each with 50 tons of clamping force; imbalance at this point is carried by a temporary support from below. The gantry's legs are moved forward so its front legs sit on top of pier 6. The front auxiliary support is then released.

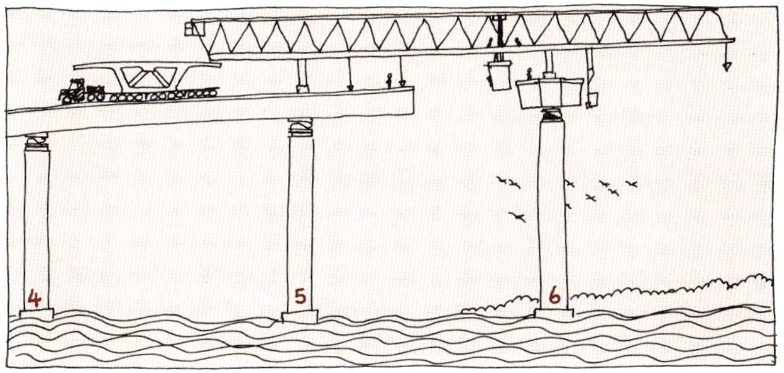

5

(5) The gantry then launches itself forward until the leading edge is halfway between piers 6 and 7. Then the segment on the other side of the pier is erected and again clamped by eight post-tensioning bars. This segment also gets temporary support from below. Permanent strand tendons are then installed to go through all three segments significantly adding to the clamping action.

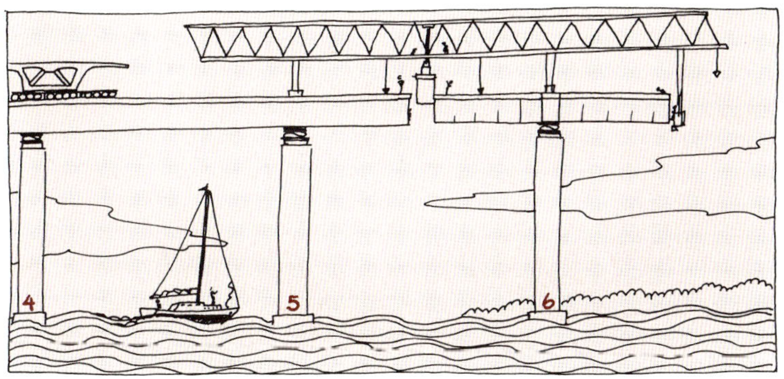

6

(6) Segment erection then continues in similar fashion adding segments in pairs, first to one side, then the other. Each time a pair is erected the strand tendons are added with the longest tendon going from mid-span between piers 5 and 6 to mid-span between piers 6 and 7. Imbalance during this phase is carried through four triangular stabilizing arms attached between the erected deck and the lower chord of the gantry. When all of the segments are placed, the closure section is formed, additional tendons in the bottom slab are tensioned and the cycle begins again.

rapidly, becoming hard in an hour or two. While the product is available in three chemical compositions for low-, medium- and high-temperature conditions (the lowest being suitable for 40 to 55 degrees), Connecticut winter temperatures typically go well below that.

The supplier first ran tests to demonstrate the performance of the epoxy down to 20 degrees F. The curing process slowed down below 40 degrees F. and stopped curing at around 20 degrees F. The curing process, however, resumed without detriment when temperatures rose.

While some thought the tests demonstrated that there really wasn't a problem at all, others were concerned about structural effects and about water entering the joints in segments with uncured epoxy — the long-term corrosion protection could not be sacrificed for erection speed.

The solution worked out was to apply heat to the joint, raising the surface and substrate temperatures to 40 degrees before applying glue. After joining two segments, electric blankets were draped over both the interior and exterior surfaces and heated until the epoxy cured. This procedure slowed segment erection a little and tended to limit erection to two segments per day. However, it did allow this work, which had now become the critical path to project completion, to continue and by Easter, when the weather started to improve, erection had progressed to pier 10 on the east bank.

In January, 1992, there was a change in leadership at the ConnDOT office in Norwich. Mike Lavallee took over when district engineer Norman Angel retired. Angel's decision to retire came suddenly and somewhat unexpectedly to his colleagues, at a time when the department was offering incentives for senior personnel to take early retirement.

Among others who left at this time were Dr. Robert Gubala, chief engineer; Walter Cox, assistant chief engineer; Dan Coffey, director of engineering; and Arthur Hourihan, director of construction, all of whom had been intimately involved in the planning, design and funding of the New Baldwin Bridge. There was a major reshuffle in nearly all ConnDOT departments and it would be several months before things settled down again to business as usual.

Tragically, Norman Angel, having served with ConnDOT for over 33 years, enjoyed barely two months of retirement before he died in April, 1992. He had taken a great interest in each stage of the new bridge and it is very sad that he did not live to see it completed.

By mid-May, the first crossing was complete and attention switched to returning the gantry to the west shore and beginning the construction of the eastbound deck segments.

There seemed to be only three options:

1. Move the gantry back under its own propulsion and slide it sideways to the eastbound deck.
2. Dismantle the gantry into sections, truck it back and reassemble it.
3. Jack it up onto trailers and haul it back in one piece.

After careful consideration and quite a bit of additional engineering by consultant Finley McNary and checks by the designer, it was agreed that option 3 was best. On a schedule basis, this was the clear winner. The other schemes would have taken several weeks longer to complete.

PPO elected to subcontract the work to Williams Crane and Rigging from Richmond, Virginia. Preparation for the rollback took about three weeks, but on the morning of May

28, 1992, the gantry was on the Old Lyme side of the river at 6:30 a.m., and by 10:00 a.m., it was back on the Old Saybrook side.

Only one small problem occurred. Just as the front end reached the Old Saybrook abutment, one of the front pivot supports slid partly off its bearing. This caused quite a bit of concern at first but some additional restraint plates were welded onto the support to prevent further slippage, and the reversal was completed without further problems.

By 9:00 p.m., the gantry was finally lined up behind the

The launching gantry gets a free ride back to Old Saybrook from two tractor units. This operation was subcontracted to Williams Rigging of Richmond, Virginia. Probably no single event was more interesting or dramatic to watch. The move from the west bank to the east bank was accomplished in less than three hours.

Margaret Thibodeau sits at the controls of the overhead gantry crane 40 feet above the bridge deck. The long vertical tube in the middle of the crane is a single cylinder hydraulic ram used to pick and place the 140-ton segments.

Nighttime shot of the gantry in January, 1992, during the crossing of the westbound span. PPO took every opportunity to make maximum use of mild weather, working long shifts to erect segments when conditions allowed the epoxy to cure without need for electric blankets.

eastbound abutment. A few days later, the gantry moved to pier 1 for a second time and started placement operations for the eastbound deck.

Placing segments on the second span

The substructure was all complete, casting of the segments would be complete in less than a month and the all-important gantry rollback was done. Was it all downhill from here on? Not quite.

The second span went much faster, of course, now that crews had perfected their tasks. Along the way they also developed many simple aids and devices to make each operation easier and more efficient.

Just after the Fourth of July weekend, the contractor elected to begin 24-hour operation, running five and sometimes six 12-hour shifts back to back until the erection was completed in late September. PPO thus assured that they would not run into the cold weather problems of the previous winter. (Hindsight proved this to be a very strategic move, since the winter of 1992-1993 was much colder than the two previous years and would have caused considerable hardship.)

"Equipment doesn't know it's working overtime," commented William Sasser, ConnDOT's on-site project engineer as he watched consecutive shifts speed the second crossing. The 24-hour operation helped get major pieces of equipment off the project sooner, which cut overhead costs.

For PPO erection engineer Elie Homsi, it would be long days — not infrequently followed by long nights when he would be called back out to address some problem that had come up during the night. Problems had to be solved quickly so work could continue. Dewey Martin, Parsons Brinckerhoff deputy project manager, commented about switching to the night shift, "Production rates for night shift operations sometimes fall short of normal daytime shift work. But here the night shift's production through the summer months often exceeded daytime operations — there was some competition between the two shifts and it worked really well. From my own perspective, I liked the night shift work. There were fewer distractions from the daily routine. You could really focus on your work and accomplish a lot."

The last segment on the eastbound bridge was placed on September 22 during the night shift. There had been some talk earlier in the week about asking local dignitaries to join workers to witness the event, but the invitations were never issued because the schedule indicated the final segment would be placed around 2:00 A.M.. There would be plenty of time for fanfare and celebration on opening day!

The last closure pour proved to be the most troublesome when a sharp drop in temperature on the night of placement resulted in some unexpected bridge shortening which had not been provided for. Part of the closure section sheared off and failed. The damaged section was sawcut, chipped out and replaced. This wasted about one week of time but the end product was uncompromised.

The gantry's work completed, it was dismantled on the east shore, each part being carefully labelled and trucked to Perini's storage facility in Bow, New Hampshire. By mid-October, the gantry was gone.

But there was still work to be done. The bridge and the approaches would not be ready to carry traffic for another seven months.

OPERATING THE LAUNCHING GANTRY

Margaret Thibodeau ran the gantry crane during daytime operations throughout the construction process. "It was a really good experience for me," she says. "As an operating engineer, I've run many types of cranes but nothing quite like this launching gantry. When I got the union's call to report to the job site, no one told me I'd be operating the gantry crane. It was a little intimidating at first, but actually, the controls were easier than other equipment I've run. Once in a while something unexpected occurred — which I guess happens with any new equipment like this — and we'd have to improvise till we got the problem sorted out. It's a great project and I'm proud to have been a part of it."

The last segment is erected during the night shift on September 22, 1992. A Christmas tree placed on the segment signifies that another milestone has been accomplished.

The under-the-bridge inspection catwalk and an inspection monorail had to be installed through the gap between the spans before median concrete could be placed. This $2 million inspection system indicates how seriously maintenance is now taken.

CHAPTER 8 Final Details

On the bridge itself, concrete parapets and medians had yet to be completed. Fences and guard rails had to be installed. Lighting, signing and a motorist emergency call box system were to be installed, pavement had to be laid and a lot of work on the approaches remained.

In fact, work had begun on the inner and outer parapets flanking the sidewalk on the westbound bridge by mid-June 1992, and was completed in about three months. Concrete placement on the south parapet of the eastbound bridge was about 50 percent complete by the time the gantry completed its second crossing.

Work on the two medians, however, could not begin because, among other things, the forms had not arrived. But even more important was the fact that only a two-inch gap would be left between the two bridges, so all work between the spans which required access from the top would have to be done before the median concrete could be placed. This work quickly became critical to the schedule.

First an inspection catwalk would be preassembled on the west approach where it would be painted and then lowered through the median gap and hung from the inner segment wings. Then rails for a tramrail hydraulic inspection device would be installed. The inspection platform is equipped with its own power system and two hydraulically operated platforms for inspectors. These can extend out from the central platform and place an inspector anywhere on the outside surface of the bridge for routine inspections. These are undertaken every two years on all highway bridges and include underwater inspection; in-depth, detailed inspections occur every 10 years.

On most bridges, this inspection work is usually performed by a truck-mounted crane called a "snooper," but the New Baldwin Bridge is so wide that this equipment would have provided only limited inspection access. So the specialized equipment was permanently installed at a cost of over $2 million — an indication of the importance placed on bridge safety today. The Bartholomew Company supplied the inspection device and the high-carbon rail support system.

A temporary work platform, measuring about 45 by 70 feet and affectionately referred to as "the dance floor," was fabricated and assembled to hang just below the middle portion of the bridge to allow subcontractor Williams Crane and

Tented enclosures and space heaters cover the median concrete placement and curing operations to allow work to proceed through the winter of 1992-1993 — which turned out to be the severest winter of the construction period.

Vic Spera, PPO labor foreman, has been on the job from the beginning. He and his crew have been involved in just about every job-site concrete placement to date. His son Joseph is a labor steward on the job; another son, Tony, is a field inspector for Parsons Brinckerhoff.

Rigging access to install the catwalk and rails for the inspection tram.

Lastly, a six-inch fire hydrant system was installed by the Tucker Company with six hydrant outlets on the bridge to enable fire fighters to respond to any fire incident on the structure. Water will be drawn directly from the eight-inch main which runs along Ferry Road.

The Old Saybrook and Old Lyme Fire Departments' trucks usually carry about 2,000 gallons of water to any automobile accident on the bridge. Without the permanently installed hydrants, a more severe situation might require a fleet of tanker trucks or several pumper trucks in series to be able to draw water from the river, which is too far below to allow direct pumping. The new system should be capable of delivering around 750 gallons per minute.

Concrete median work did not get under way until shortly before Christmas — just as cold weather started to move in again. Luckily, most of the painting of the inspection catwalk was completed by this time. Final paint touch-up would have to wait until spring.

The contractor was now faced with a choice. Option 1 was to continue working through the winter with loss of efficiency due to the added burden (and cost) of providing shelters for the work, extra curing blankets, propane heaters, etc. Option 2 was to shut down virtually all work and come back in March when the weather improved.

ConnDOT, however, was concerned that option 2 could prevent the bridge from opening in time for summer, 1993, holiday traffic even though the project was a good nine months ahead of schedule. Discussions took place which resulted in an agreement being reached on January 6, 1993, calling for the contractor to commit to an opening date of on or before 9:00 a.m. on May 25, 1993. ConnDOT agreed to offset, in part, the contractor's additional costs and risks associated with this accelerated program and for working through the winter. To ensure the contractor's commitment, the agreement also provided that deductions would apply for each day following May 25 that the bridge was not open.

The race was on

Rumors of a "big" celebration for official opening day began to circulate on the job — things would move fast from here on. A new schedule was drawn up; everyone reviewed and discussed possible ways to expedite the remaining work — particularly the approaches. This became even more essential since PPO had relieved Manafort Brothers and Middlesex Corporation of portions of their remaining contracts on the approaches in order to better sequence the final stages of approach construction.

The remaining work included completion of the median operations, the west approach concrete slab, the installation of the four expansion joints, completion of all electrical work and the emergency call-box system, the construction of a new on-ramp from Route 156 in Old Lyme and final paving of the bridge and the two approaches. Due to the necessary sequence of operations and restriction on lane closures, it would be necessary that all final pavement on I-95 would be done at nighttime when the traffic was lightest. It was expected that this work would extend into July and possibly August.

At the time this book went to press, the contractor was

EXPANSION JOINTS

Expansion devices are notoriously one of the biggest maintenance headaches on older bridges. On newer structures, expansion joints are reduced to the minimum number possible.

There are only four joints on the New Baldwin Bridge; they're located at the abutments on each end of each span. These joints are capable of accommodating up to 18 inches of movement on the Old Saybrook side and 14 inches on the Old Lyme side.

This is not, as one visitor questioned, because Old Saybrook is any more seasonal than Old Lyme. It is because of the layout of the piers and the bearings, some of which allow sliding action and some of which are fixed. Because of the underlying geology and the substructure design, most of the sliding piers are on the west side of the river, hence the movement required is greater.

Actually, not all of that range is due to temperature variations, which account for around nine inches of the 18 inches at the Old Saybrook abutment, corresponding to a temperature range of 85 degrees F. maximum and 5 degrees F. minimum. The remainder of the movement is to accommodate permanent long-term inelastic deformation of the concrete in the bridge due to shrinkage and creep, both of which tend to shorten the structure over time. The modular joints on the Baldwin Bridge supplied by Watson Bowman are considered state-of-the-art for long-span bridges.

ahead of schedule and, barring the remote possibility of major problems which may arise over the next few months, it is expected that three lanes of the new bridge will be opened to traffic on or before May 25, 1993, and the project will be fully completed by the end of summer.

"The quality of the construction has been first class throughout," comments Mike Lavallee, ConnDOT district engineer for District II in Norwich. "As with any contract, there have been difficulties along the way. This contract is a big one, so any dispute tends to be significant. There has, however, been a real attempt to resolve these differences as they arise and so far, we've managed pretty well. It has been a real partnering approach."

Demolition of the old Baldwin Bridge

As traffic is routed onto the new bridge and as the last few items of work are completed, the contract to demolish and remove the existing structure will be advertised. This work will probably begin in the fall of 1993 and will take 12 to 18 months to complete.

To develop plans and specifications for the demolition of

The bridge nearing completion displays strong sweeping lines against the Connecticut skyline. The low profile should more than satisfy those who feared a larger bridge would detract from the scenic beauty of the lower Connecticut River. Sharp observers may notice that the roadway of the new bridge has a more gradual rise than the old bridge, providing motorists with a longer sight distance as mandated by today's codes.

the old bridge, ConnDOT again turned to Parsons Brinckerhoff. The specifications call for the bridge to be completely removed down to approximately six feet below river bottom, eliminating any possible hazard to navigation. A similar project on the Charter Oak Bridge further upstream at Hartford began in May 1992, and is ongoing. The old Charter Oak Bridge is in fact quite similar to the old Baldwin Bridge and a similar sequence for demolition will likely take place.

In any event, the same environmentally conscious restrictions required in the new bridge construction will again be required for the removal of the old bridge to be sure that Connecticut's most important watercourse suffers no pollution as a consequence of demolition.

No construction debris will be allowed to fall into the river. Cofferdams will again be required for lower pier and foundation removal, and blasting will be permitted only lower than six feet above mean sea level. Restrictions for dredging and pile driving will again be enforced to minimize effects on migrating fish.

Of interest to boaters and fishermen, and as a significant benefit to the community, tied into the demolition contract will be construction of a new boat launch facility and a 324-foot fishing pier. The facility will be much larger than the existing boat ramp with greater capacity and more parking space. A handicapped ramp will also be included and toilet facilities, picnic tables and landscaping will complete the site. It is expected that this work will be undertaken concurrent with the bridge demolition, but provisions are included in the contract that a boat ramp and parking facilities must remain open at all times during construction.

A look to the future

So the story of the crossings is now complete — from dugout canoes, driftwood rafts and colonial ferries through the construction of three bridges. An interesting observation might be made by someone fascinated by numerical progressions as follows:

- The 1911 bridge cost $475,000 and was used for 37 years.
- The 1948 bridge cost $5.5 million (or approximately 10 times more than the one it replaced) and was used for 45 years.
- The 1993 bridge cost about $100 million (or approximately 20 times more than the one it replaced) and mathematically might be expected to last 53 years.
- The next bridge might then be built in the year 2046 and, if the progression holds true, may well cost a staggering 30 times more than today's bridge, or $3 billion.

Perhaps by the year 2046 ground transport will be a thing of the past or perhaps the politicians in Washington will have better learned to control inflation so that the cost escalation in the $3 billion range is brought back to something more reasonable. Who knows?

What's certain is that this bridge has been designed for a minimum of 50 years design life and could quite likely last twice as long. Many features have been included, like the requirement for 100 percent epoxy-coated rebar and the high strength superstructure concrete, which should prolong the useful life of the structure.

Routine maintenance on a properly constructed concrete bridge is normally less than on a steel bridge. There is no

Tim Delicia (left), labor superintendent, and Roger Oakley (right), carpenter superintendent, discuss arrangements for the day's concrete placements.

Parsons Brinckerhoff Senior Inspector Leroy Conners has been on the job from the beginning. He joined the company as an inspector in 1969 and has worked on numerous projects including the Sunshine Skyway Bridge and the Newport Suspension Bridge. Often leaving for the next job before opening day, he plans to be at the New Baldwin Bridge for ribbon cutting.

expensive paint removal and recoating required. The concrete of the Baldwin Bridge piers is coated with a protective compound which penetrates the upper surfaces providing a barrier to moisture and chemical attack, which over long periods of time can cause deterioration of the concrete and corrosion of the reinforcing or post-tensioning steel.

Many bridges, such as the two predecessors of the new bridge, have had to be replaced before they had fulfilled their design life. The reason for early replacement had more to do with higher-than-anticipated traffic volumes than it did with premature deterioration.

Some might ask why this did not happen in the case of the Brooklyn Bridge, which celebrated its centennial in 1983, particularly as it now carries much heavier traffic than the horse and buggy and mini-trains on which John Roebling based his design. The answer lies partly in the huge safety factors (or reserve strength) designed into this remarkable structure, probably at least six times greater than was thought necessary.

Today, loadings and strength of materials are much better understood and extensive bridge design codes exist with which designers must comply. In fact, these do provide a high margin of safety with respect to all possible load combinations which might be expected, including dynamic effects from wind — a field that Roebling probably knew the least about. But who can accurately predict future traffic volume or trends which are more likely to be a problem for traffic movement than structural loading? Anyone who lives in New York knows that free-flowing traffic on the Brooklyn Bridge is not the norm — or on any other bridge in New York for that matter.

THE 1984 RENEWAL PROGRAM COMES TO AN END

Current Commissioner of Transportation Emil Frankel recently announced that ConnDOT is expected to spend up to $650 million in projects during the years 1993 to 1995. He said that this capital program will bring to a substantial conclusion the infrastructure renewal program launched by the state in 1984.

That program, he said, is remedying years of infrastructure neglect which threatened to impede the movement of goods and services from their providers in Connecticut to their markets worldwide. It is an aggressive, standard-setting program which recognizes that a superb infrastructure is an essential prerequisite for economic development and job creation.

Jay Davidson, project manager for PPO, and Scott McNary of Finley-McNary, were both honored by *Engineering News Record* for their outstanding achievements in the successful construction of the Baldwin Bridge. Jay holds one of the plaques they received at New York's Plaza Hotel on February 10, 1993. Accepting the award, Jay commented, "It's nice to be recognized for your efforts on a job like this — but this is not an individual award, its a team award and I'd like to thank everyone who helped make this project a success."

Not too well known is the fact that the Brooklyn Bridge went through extensive modifications around 1948 to accommodate increases in highway traffic at a cost of over $9 million. Now, of course, the Brooklyn Bridge is a national historic landmark, deservedly so, and will likely be preserved at whatever cost for a very long time.

Chances are the New Baldwin Bridge will never be accorded such stature. It should, however, have considerable reserve capacity well beyond the traffic volumes used in the design. These were based on predictions for the year 2010 of 38,800 vehicles a day in each direction.

Of the four lanes on the new bridge, the inner two lanes are expected to carry strictly I-95 through traffic. One lane will be for local merging or exiting traffic and the other lane is intended for strictly local traffic. The protective fence with limited through visibility will discourage some of the tendency to slow down to admire the spectacular views.

Summer 1993 traffic should be somewhat alleviated by a minimum of three lanes in each direction on the new bridge. However, it will be summer 1994 before the traffic engineers and the general public truly know for sure if the bottleneck problem is a thing of the past. Let's hope so.

As for the bridge builders, many of them have already moved on. Bridges of this size, which attract that strange breed of men and women who are drawn to build them, are few and far between. The builders seldom find themselves in one place for too long. It is a demanding life, but a rewarding one. The New Baldwin Bridge will stand as testimony to the dedication of the bridge builders long into the 21st century. As English writer and critic John Ruskin once said, "When we build, let us think that we build forever."

KEY DATES

Start construction	02-01-90	
Dredging	02-13-90	- 03-31-90
Test piles	04-20-90 (pier 1)	- 10-10-90 (pier 10)
Production piles	06-08-90 (pier 1)	- 12-20-90 (pier 10)
Cofferdam installation	02-14-90 (pier 3)	- 02-22-91 (pier 9)
Tremie placements	08-21-90 (pier 3)	- 07-31-91 (pier 9)
Foundation placements	08-22-90 (pier 1)	- 09-26-91 (pier 9)
Pier construction	09-10-90 (pier 1)	- 02-20-92 (pier 9)
Precast operations	11-13-90	- 09-09-92
Launch gantry assembly	04-23-91	- 07-22-91
Erection operations	07-13-91 (pier 1)	- 09-22-92 (span 10)
Parapet work	06-17-92	- Est. 05-93
Fixed inspection (catwalk) installation	09-01-92	- Est. 04-93
Moving inspection device installation	11-11-92	- Est. 04-93
Hydrant system installation	01-04-93	- Est. 04-93
Bridge open to traffic	Est. 05-25-93	
Project completion	Est. 08-93	- Est. 03-95
Demolition contract	Est. 10-93	
Fence/bridge rail installation	11-10-92	- Est. 05-93
Pavement	05-93	

A FEW FACTS AND FIGURES

Tremie concrete (piers 2 - 9)	15,573 C.Y.
Foundation and pier bases concrete (pier 1 to pier 10)	34,110 C.Y.
Pier columns concrete	15,780 C.Y.
Superstructure concrete (precast)	32,248 C.Y.
Superstructure concrete cast-in-place	773 C.Y.
Number of precast segments	488
Parapet and sidewalk concrete	2,450 C.Y.
Granite masonry (piers 2 - 9)	732 C.Y.
Rebar	12,725,500 lbs.
Post-tensioning (strand)	3,045,768 lbs.
(bars)	193,726 lbs.
Excavation earth inside cofferdams	27,937 C.Y.
Piling (all HP 14 X 89)	15,814,809 lbs.
Biggest tremie seal (pier 9)	2,860 C.Y.
Biggest footing	2,112 C.Y.
Total no. of piles (piers 1 to 8 and 10)	1,776
Most piles in a pier (pier 6)	328
Caissons in pier 9 (HP 14 X 550) (each in 30" dia. drilled caisson shaft)	28
Gallons of epoxy adhesive (approx 10 gallons per joint)	5,000 gallons
No. of subcontractors	40
No. of PPO employees - max-summer, 1991	185
- total to date	433
No. of PBCS/L-C employees - max-summer, 1991	22

Trades involved:

carpenters - Local No. 24
ironworkers - Local No. 424
laborers - Local No. 547
operating engineers - Local No. 478
concrete masons - Local No. 1, 2, 12 & 16

Bridge Builders

List of Subcontractors (individuals not listed)

A & W ARTESIAN WELL CO.
ADVANCE STEEL
ALL-STATE SILT FENCE CO.
ALBINI CORP.
AUCIELLO IRONWORKS, INC.
BLAKESLEE, ARPAIA, CHAPMAN, INC.
CARROCCIO & COVILL ASSOC.
CONCRETE HYDRO
CONKLIN & SOROKA
CONNECTICUT FENCEMEN
CONNECTICUT LIGHT & POWER
CONNECTICUT WATER CO., THE
COTE & DAUGHTERS ELECTRIC
DESCO PRODUCTS OF CT, INC.
DUCCI ELECTRIC CO., INC.
GATES & WHALING CITY, JV
GUARDRAIL, INC.
JACOR, INC.
LAL CONSTRUCTION CO., INC.
LAMAY PAVING CO., INC.
LAUGENI A. & SONS, INC.
MCCLURE CONSTR. CO., INC.
NORTHEAST UTILITIES
NORTHEASTERN CLEARING, INC.
PARKWAY CONSTRUCTION, INC.
PASQUALINI CONSTR., INC.
PREPAKT CONCRETE CO., THE
SAFETY MARKINGS
SANTORO SEAL COATING
SCHUMACK CONSTR. CORP.
SEA-JAY MARINE CONSTR.
STATE RAIL & FABRICATING CO.
STRONGWORK DIVING, INC.
TILCON CONN., INC.
TUCKER COMPANY, THE
WATERS CONSTRUCTION CO., INC.
WILLIAMS CRANE & RIGGING, INC.

Connecticut Department of Transportation

ANGEL, NORMAN
AVINO, VINCENT
BECKWITH, MILTON
BISHOP, WAYNE
BUTZGY, ARTHUR
CABELUS, ROBIN
CHAMBERLIN, WILLIAM
CIRIELLO, ALPHONSE
CONTI, JOSEPH
COX, WALTER
CROUTHAMEL, PHIL
DEVINE, JOANNE
DOUGHTY, JAMES
FARGNOLI, KENNETH
FINNEMORE, MICHAEL
FRANKEL, EMIL
GAGE, STEVEN
GIRASOLI, CARL
GRUHN, ARTHUR
GUBALA, ROBERT
GUMACHE, ROBERT
GUZZO, DAN
HOURIHAN, ARTHUR
HUZAR, PETER
KASUGA, MICHAEL
KLOSKOWSKI, ROBERT
LAWRENCE, KEITH
MULLIN, NEIL
NELSON, CARL
ORGOZALEK, RAY
O'BRIEN, JOHN
PANTELEAKOS, CHARLES
REDAKAS, KEVIN
SASSER, WILLIAM
SIMONEAU, RICHARD
STARK, WILLIAM
STICHT, WILLIAM
STRENKOWSKI, DENISE
SULLIVAN, JAMES
TEFT, CHARLES
ZYSKOWSKY, STAN

Federal Highways Washington Administration

BARAKOS, JAMES
BRIGGS, DAVID
CHILSTROM, JOSEPH
DIRKS, ROBERT
PRICE, DAVID
RAMIREZ, ROBERT

L-C Associates

BERGERON, JIM
CHIN, HOWARD
CHUANG, FRANK
CLARKE, BARRY
DOANE, WILLIAM
FLYNN, DAVID
HAINES, GEORGE
MARTUCCI, JOHN
MIKULAK, PAUL
MULLINS, ROGER
PATEL, DIPAK
RITTER, DOUGLAS
ROWE, PETER
SLATER, JON
WRIGHT, RANDALL

Parsons Brinckerhoff Construction Services

BROWN, KATHLEEN
CHAKKAPHAK, POMSIT
CONNERS, LEROY
COOPER, TIMOTHY
DAVIS, SCOTT
FALARDEAU, DONALD
GILL, PATRICK
GREEN, DONALD
GROMKO, TONI MARIE
HOYT, PAUL
HUBBARD, IAN
JOHNSON, BEVERLY
KINGSTON, DAVID
LONG, LORI
MAJEWSKI, STEVEN
MARGIOTTA, ANTHONY
MARTIN, DEWEY
MARTIN, TARISHIA
NELSON, PAMELA
ORRISON, JOHN
QUARANTA, THOMAS
RESEIGH, CHRISTOPHER
SMITH, GERALD
SOFIA, MARK
SPERA, ANTONIO
TONG, DAVID
TOWELL, PAUL

Parsons Brinckerhoff Quade & Douglas

ABRAHAMS, MICHAEL
ALLEN, JOHN
BARTOLETTI, JOHN
BRACKEN, WILLIAM
EGEMAN, AYNA
HILLMAN, JOHN
LASSITER, GREG
LAWRENCE, EDMUND
LEVY, MORRIS
MEDINA, HECTOR
MURILLO, JUAN
PHILLIPS, JAMES
REJCHA, CHARLES
SHOWERS, JOSEPH
SILANO, LOU
WALTMAN, GERALD

Perini/PCL/O&G Joint Venture

ADAMOSKI, JOHN
ALDERMAN, ELLIOTT
ALICE, LAWRENCE
ANDERSON, ROBERT
ANDREWS, JULIAN
ANDREWS, KEVIN

APPLEBY JR., CHARLES
ARNOLD, TIMOTHY
ARPIN, MICHAEL
ARRIGONE, JON
AUCLAIR JR., ARTHUR
AUGUSTINE, THEODORE
AVERY, BARRY
AVERY, ROBERT
AWEZEC, SUE
BACHANT, CHRISTOPHER
BACHRACH, DENISE
BACON, CHRISTOPHER
BACON, TODD
BANKS, RICHARD
BARBOZA, BENJAMIN
BARRICELLI, DIEGO
BARROWS, PAUL
BEEBE, SCOTT
BERNSON, ROBERT
BERRY, JOYCE
BERRY, MAURICE
BERTOLINI JR., RODY
BINGHAM, GARY
BLANCHET, CHRISTOPHER
BLIVEN, GEORGE
BORNYAK, MICHAEL
BOUCHARD, THOMAS
BOUDREAU, DENIS
BOYLAN, S. WILLIAM
BRAYLEY, ALAN
BRAYNE, KEITH
BRAYNE, PAUL
BREWER, GARY
BROCCOLI JR., LEROY
BROOKS, THOMAS
BROWN, EARL
BROWN, EDWARD
BROWN, EUGENE
BROWN, LOUIS
BROWN, MICHAEL
BUICE, WILLIAM
BURELLE JR., RICHARD
BUSEY, MICHAEL
BUSH, EDWARD
BYINGTON, RALPH
CALICCHIO JR., ANDREW
CANNAMELA, THOMAS
CANNATA, BRYAN
CARREIRO, JOHN
CASEY, ANDREW
CERRUTI, JERRY
CERVANTES, HUGO
CHASE, STEPHEN
CHUMPITAZI, ALEX
CICARELLI, PASQUALE
CICCARELLO, ANGELO
CLARK, WILLIAM
CLEFSTAD, JOHN
CLOSE, EDWARD
COBURN, GARY
CODERRE, ALFRED
COLLELO, DAVID
COLLELO, JAMES
COLLELO, PETER
COLLELO, ROBERT
COLLELO, VINCENT
CONABLE, MARGARET
CONGDON, DAVID
CONTINO, RAYMOND
COVIN, ANTHONY
CRANDALL, KEVIN
CULP, STEVEN
CUNNINGHAM, PERRY
CYR, RICHARD
DALY, JOHN
D'AMBROSE, EUGENE
DAVIDSON, DONNA
DAVIDSON, HAROLD
DEASON, WILLIAM
DELMONACO, MICHAEL
DELUCIA, MARCO
DEOJAY, ROGER
DEPINA, ALFREDO
DEPRIMO, SALVATORE
DEROSA, PHILIP
DETORO, STEVEN
DEVEAU, BERNARD
DHINGRA, SUBHASH
DIAZ HOMSI, CLAUDIA
DIEDERICH, THOMAS
DILISCIA, EUTIMIO
DION, ROBERT
DIPALMA, JOSEPH
DIPALMA, MONTE
D'OSTILLO, DAVID
DOWLING, DION
DUERR, SCOTT
DUGAS, ANDRE
DUZANT, MICHAEL
EIMUTUS, KEVIN
ENGEL, GREG
ESPOSITO, ANTHONY
FACAS, ABDO
FELICIANO, THOMAS
FERRUCCI JR., ARTHUR
FESTA, CHARLES
FIDRYCH, PETER
FINE, PAUL
FIONDELLA, FILIPPO
FITCH JR., ROBERT
FITZER, JAMES
FOLEY, PATRICK
FONTAINE JR., RICHARD
FORDE, VICTOR
FOSTER, RONALD
FOSTER, YVETTE
FRANCO, RALPH
FRYKLUND, GEORGE
FULLWOOD, DWIGHT
GALLO, LOUIS
GAMBARDELLA, JOSEPH
GAMBLE, STEVEL
GANCI, ANTHONY
GARDEGA, KENNETH
GARDINER, RICHARD
GATES, ROBERT
GAUDETTE, RAYMOND
GAUDIO JR., SALVATORE
GEER, MARVIN
GENOVESE, JOSEPH
GIBSON, GREGORY
GISH, ROBERT
GNITZCAVICH, MICHAEL
GOARDINO, CARMINE
GODREAU, LLOYD
GOGGIN, JAMES
GOLDEN, THOMAS
GOLNIK, DEAN
GOMEZ SR., JOSEPH
GOMEZ, JOSEPH
GOMEZ, SHANE
GOTT, MICHAEL
GOULD, RICHARD
GRANT, GERALD
GRAY, BENNIE
GREENE, PHYLLIS
GREGORASKOS, GEORGIOS
GRETO, VINCENT
GRIGORASKOS, IONNIS
GROEPER JR., LESTER
GROSS, GENE
GUILD, ROGER
GUILD, ROY
HAJEK, PHILIP
HALL, JOHN
HAMILTON, JANE
HANNON JR., WILLIAM
HARAKALY III, STEPHEN
HARAKALY SR., THEODORE
HARAKALY, MITCHELL
HARRIS, CHRIS
HARVEY, JONATHAN
HATJE, PAUL
HAWLEY JR., CHARLES
HEATH, FREDERICK
HENDERSON, EDWARD
HENK, RONALD
HIGGANS, TILDEN
HINZE, DAVID
HIRST, EDWARD
HOMSI, ELIE HABIB
HURLBURT, SUMNER
HURLEY, WILLIAM
HURST, HOSEA
IRIZARRY, CARL JUAN
JACKSON, WILLIAM
JACQUES, RANDY
JAKUBIAK, RICHARD
JENSEN, ROBERT
JOHNSON, JOSEPH
JOHNSON, LEE
JOHNSON, MARK
JOHNSON, RONALD
JONES JR., MERVALE
JONES, DONALD ROBERT
JONES, JOHNNIE
JUDD, MICHAEL
KALISZ, MICHAEL
KENN, STEPHEN
KENZIOR, PAUL
KEVITT, CHARLES
KICHLINE, JAMES
KIGGANS JR., RAEFORD
KIGGANS SR., RAEFORD
KING, CHARLES
KIRK, THEODORE
KLINGMAN, KENNETH
KOPECINSKI, MICHAEL
KURISOO, JAAN
LABONSKY, BENJAMIN
LALUMIERE, DONALD
LAMARCO, WILLIAM
LAMARK, SAMUEL
LAMOTHE, CHARLES
LANDRY, LUKE
LANGDON JR., JOSEPH
LAPIER, CARL
LAPLANTE, DIANNE
LAVIN, CAROL
LAVOIE JR., JOHN
LAWSON, RODNEY
LEBLANC, SCOTT
LEBOVITZ, MINDY
LEFRANCOIS, EDWARD
LENIO, ERIC
LENIO, JOHN
LENIO, KURT
LEONARD SR., JAMES
LEVERONE, DAVID
LIENKE, PAUL
LIPPINCOTT JR., RICHARD
LOVELY, WILLIAM JOHN
LUCATINO, ALDO
LUCIANI, FRANK
LULAY, KARL
LUSA, RICO
LUZZI, BRANDON
MACCALMONT, JOHN
MACHOLD, MARTIN
MACLEAN, THOMAS

MADDOCKS, PETER
MAGLIARI, SALVATORE
MAILETT, MARK
MAIOLO, SALVATORE
MALCARNE, JOHN
MALINOWSKY JR., HENRY
MALINOWSKY JR., ROBERT
MALINOWSKY SR., HENRY
MALINOWSKY SR., ROBERT
MANISCALCO, MICHAEL
MARCINKIEWICZ, JAMES
MARINO JR., MICHAEL
MARLOWE, RICHARD
MATHESON, DAVID
MATIAS, JOSE
MATIAS, JOSE
MAUST, RONALD
MAYNARD JR., ROBERT
MCCARTHY, JOSEPH
MCCORMACK III,RICHARD
MCCORMACK II., RICHARD
MCGARRY, GEORGE
MCINERNY, PATRICK
MELILLO, ROBERT
MELILLO, ROBERT
MENDEZ, JAVIER
MICHEL, ROBERT
MICHEL, ROBERT
MILDRUM, CHRISTOPHER
MILESKI, MITCHELL
MILLS, ROBERT
MOELLER, DANIEL
MOELLER, DANIEL
MONACO, ROBERT
MONGILLO, FRANCIS
MONTONE, COSTABILE
MONTONE, MICHAEL
MORGAN JR., LEOLIN
MORGAN, LEOLIN
MORGAN, RANDY
MORGAN, ROGER
MORIN, DENNIS
MORSE, SHERRY
MOSES, TERRY
NAPARTY, KATHLEEN
NAPOLITANO, JAMES
NAVARRO, DAVID
NEGRELLI, AMEDEO
NIEVES, MANUEL
NI, HUANG
OAKLEY, ROGER
OJEDA, ANGEL
OJEDA, HOLGER
ONEGLIA, RAYMOND
ONUFER JR., JOHN
O'PANEL, FRANCES
ORSOS, NANDOR
PAGNANO, THOMAS
PAIGE, STANLEY
PALMER, KEVIN
PALMIERI, DOMENICK
PAPA, FRANK
PAPA,MICHAEL
PAPROCKI, JOHN
PARADIS, RENA
PARKER, FRANK
PARRISH JR., ANDREW
PARRISH JR., ANDREW
PATERNO, VINCENZO
PAUL JR., EDWARD
PAUL, DAVID
PEAVY, RICHARD
PECKA, ADOLPH
PEEL, PETER
PELLEGRINO, JOSEPH
PELLETIER, ANDREW
PENDLETON, JAMES
PERKINS SR., RUSSELL
PERKINS, RAYMOND
PERKOWSKI, WILLIAM
PERRY JR., LOWIS
PESCATELLO, LYNN
PHILLIPS, ROBERT
PICARD, RONALD
PILIPKOWSKI, JOSEPH
PISCITELLI, PATRICIA
PRAY JR., JAMES
PRAY SR., JAMES
PREVITI, KATHLEEN
PRISHWALKO, JOCELYN
PROTO, GELL
PUBLISI JR., JOSEPH
RAINVILLE, PAUL
REAGAN, GREGORY
REED, ALBERT
REILLY SR., JERRY
REYNOLDS, ARTHUR
RICHARDSON, MITCHELL
RIFFLE, GERALD
RILEY, DOUGLAS
RILEY, THEODORE
RITACCO, FRANCESCO
RITCHOTTE, WILLIAM
RIVERA, JEHUDI
RIVERA, JORGE
ROBINSON, GENE
ROBINSON, PATRICIA
ROCKWELL, DWAYNE
RODRIGUES, ALIRO
RODRIGUEZ, EDISON
RODRIGUEZ, RAMON
ROGERS, DENNIS
ROMANOFSKI, ANDREW
ROMAN, WILLIAM
ROOT, EDWARD
ROSSELLI, PHILIP
ROYSTER, EARLE
RUDEIGER, CARL
SAAD, JOSEPH
SACZYNSKI, MICHAEL
SANDERS, DONALD
SANTOS, MATTHEW
SAPIA, SALVATORE
SARKEIT, LORI
SAUERESSIG, JOSEPH
SAUERESSIG, ROBERT
SAVALLE JR., VINCENT
SAVINO, FRANK
SCHERP JR., ARTHUR
SCHNURBUSH, DANIEL
SCHULL, JEFFERY
SCHULL, WILLIAM
SCHULZE, MICHAEL
SCHUSSELL, RICHARD
SEARS, MICHAEL
SHAWCHUK, MICHAEL
SHERBURNE, DEWEY
SICIGNANO JR., JOSEPH
SIDIROPOULOS, KONSTANTINOS
SIGERSMITH, GERHURDT
SIKORSKI, EDWARD
SILVA JR., ROBERT
SILVA, JOHN
SILVA, RICHARD
SILVA, ROBERT
SLANE, THOMAS
SLIPPERLY, STEVEN
SMITH JR., JOHN
SMITH, GARY
SMITH, JAMES
SNURKOWSKI, MICHAEL
SODEN, RICHARD
SOWELL JR., DUBIE
SPERA, JOSEPH
SPERA, VITTORIO
SPERA, VITTORIO
STACK, THEODORE
STAMOS, JOHN
STAVESKI, CRAIG
STEELE, WILLIAM
STEPHEN, DANIEL
STODDARD, JAMES
STONE, VALMOS
SULLIVAN, MARK
TALAMONA, MICHAEL
TARANTINO, ANDREA
TARQUINO, JEFFREY
TAYLOR, JOHN
THIBODEAU, MARGARET
THOMASSON, CHARLES
THOMAS, MAHLON
THOMAS, SYLVESTER
THOMPSON, TODD
TURNER, K.F.
TYNDALL, JOSEPH
VALENTA, JOSEPH
VANACE, HARRISON
VANWART, MARK
VERONICK, LEON
VICHAS, WALTER
VONDECK JR., BURTON
VOSHEL, CARNELL
WALTERS, WILLIAM
WARD, RAY
WEEKS, LISA
WHITE JR., FRANK
WHITE SR., DANIEL
WHITE, GEORGE
WICKLEIN, CHARLES
WILLIAMS, JAY
WILLIAMS, SYLVESTER
WINTON IV., WILLIAM
WOMACK, WILLIAM
WOODS, HARRY
WORST, VICTOR
YARRINGTON, HAROLD
YATES, SHERMAN
YESINOWSKI, DAVID
YURKOWSKI, BRIAN
ZIMMERMAN, WILLIAM
ZISK, KATHY

Major Suppliers

AUCIELLO IRON WORKS
BARKER STEEL
BARTHOLOMEW COMPANY
BETHLEHEM STEEL
CANRON
CARBOLINE
COLONIAL IRON WORKS
J.O. DOUGALL
DURASTONE
FLETCHER GRANITE
PRECISE FABRICATING
SIKA CORP.
STANDARD STRUCTURAL
SYRACUSE RIGGING
TILCON
UNIDYNE
VSL CORP.
WATSON BOWMAN

Engineering Consultants

BRYANT ASSOCIATES, INC.
FINLEY McNARY ENGINEERS, INC.
GOLDBERG ZOINO & ASSOCIATES, INC.
MUESER RUTLEDGE CONSULTING ENGINEERS, INC.
REMY P. PAPP CONSTRUCTION CONSULTANTS, INC.

Great care has been taken to list everyone accurately. We deeply regret any errors or omissions.

Glossary

Abutment
The masonry at each end of a bridge supporting the end span and often acting as a retaining wall for the earth embankment behind it.

Air lift
A device used for excavation of those places in cofferdams that are inaccessible to the clam bucket. It pumps presuruzed air to the bottom of an open discharge pipe putting surrounding riverbottom material in suspension. Pumps then pull the debris-laden water up the pipe and discharge it into a barge.

Anchor plate
A plate fastened to a structure through the center used to lock the post-tensioning forces in place.

Ashlar
A wall of squared building stones cut true on all faces to allow for very thin mortar joints.

Bearing plate
A steel plate component of the disc bearings supporting the pier and abutment segments transmitting the loads to the pier cap below.

Caisson (drilled)
A cylindrical hole drilled into the rock, into which a heavy pile section is placed and fixed permanently by underwater grouting.

Caisson (pneumatic)
An air and watertight chamber or shaft within a foundation, in which workers excavate material below water level.

Cantilever
A structure supported at one end, reaching over space. Two of these projections, coming from opposite piers, meet to form the span of a bridge.

Cap
The top part of a column or pier, supporting bridge bearings and the superstructure.

Chord (upper and lower)
The top and bottom portions of a truss.

Cofferdam
A temporary watertight enclosure built around a foundation that is under water or in waterlogged soil. They are used for excavation, construction or repairs.

ConnDOT
Connecticut Department of Transportation

Deck slab
The top section of the concrete box girder on which the roadway surface is placed and on which traffic will ride.

Disc bearings
Structural assemblies which include a hard neoprene disc between steel plates. They form the interface between the bridge deck (superstructure) and the piers (substructure) and allow the structure to "breathe" under temperature changes and long-term deformation.

Earthquake restrainer
A device usually of steel or concrete (or both) incorporated into a structure to restrict movement in the event of an earthquake.

Gantry
A horizontal crane with a long framework (or truss) that has multiple supports for supporting an overhead travelling crane used for the erection of the precast segments.

Grillage
Horizontal or crossing beams, usually of steel or timber, for distributing the load of a structure over its footing.

Laitance
A cement-based mud in the top six to twelve inches of an underwater concrete pour which is much softer and waterlogged than the rest and is normally removed prior to further concrete placement.

Lift
The concrete placed between two consecutive horizontal construction joints, usually containing several layers or courses.

Match cast segment
In a precast segmental bridge each segment (except for pier segments) is made using the previously cast segment as the vertical form for one side and a bulkhead for the other. The previously cast segment used in this way is called the match cast segment.

O & G
O & G Industries, Torrington, Connecticut

PCL
PCL Constructors, Inc., Edmonton, Canada

Perini
Perini Corporation, Framingham, Massachusetts

Pier
A solid column supporting the deck of a bridge.

Pile
A long, thin structural member made of timber, concrete or steel and embedded in the ground in groups to carry the load through soft material at the riverbed level, sometimes extending all the way to bedrock.

Post-tensioning
The precompression of reinforced concrete by tensioning steel tendons through embedded ducts after the concrete has hardened.
PPO Joint venture of Perini, PCL and O & G.

Prestressing
Same as post-tensioning above, except in this case the tendons are tensioned before the concrete is poured, the strands then being directly in contact with the concrete placed around them.

Rebar
Steel bars placed in concrete for reinforcement to improve the ability to resist tension.

Rip rap
Pieces of stone, usually weighing from 15 to 150 pounds each, used for foundations or revetments against the force of water.

River scour
The erosion and movement of sand and other material at the bottom of a river by the force of flowing water. In extreme cases, scour can undermine bridge foundations and cause them to shift.

Rocker bearing
A type of bridge support not commonly used today but used on the old Baldwin Bridge. This type of bearing rotates to allow deck movements but cannot move horizontally, unless carried on rollers.

Segmental
A type of structure formed in sections or units which can be precast and assembled on the job or cast in place.

Sheet Pile
A steel pile or sheeting that interlocks to form the walls of a cofferdam.

Single-cell box
A type of bridge section which has the shape of a closed box with a single internal void.

Sloping webs
Webs are elements of the box girder which connect the top deck to the bottom deck. They can be vertical or sloping.

Spine beam
The central beam which runs down the middle of the box section which adds stiffness to the upper deck section.

Stringers
Horizontal structural beams which, in a steel bridge, normally support the bridge deck and carry their load to the main or primary beams and from there to the support piers.

Struts
Internal concrete braces designed to stiffen the concrete box and provide support to the box section between the span from web to web.

Substructure
The foundation or lower part of a structure which supports the top part or superstructure.

Template
A pattern of wood or metal that guides and ensures accurate placement of piles or other objects.

Tendon
A steel member made of seven-wire strands or high strength bars which is tensioned to provide precompression of the concrete in a post-tensioned bridge.

Tendon ducts
Pipes of plastic or galvanized steel cast into a post-tensioned structure through which steel tendons are installed and tensioned.

Tremie Pipe
A pipe and hopper through which concrete is poured under water so that the concrete does not mix with the water. The concrete is deposited in the hopper and flows to the bottom. The entire apparatus can be lifted by a crane.

Truss
A framework made up of a top chord, a bottom chord and diagonal bracing, which when assembled has much greater strength and stiffness than the individual components.

Index

About the Author

IAN HUBBARD was born in Glasgow, Scotland, June 1952. He obtained a BSc in Civil Engineering at Glasgow University in 1975 and then worked in Yorkshire, England, for five years on several highway and bridge projects before moving to the United States in 1980. Since then, he has been involved in the construction of a number of major bridges across the United States including the Seven Mile Bridge in the Florida Keys; the East Huntington Bridge in West Virginia; the Sunshine Skyway Bridge in Tampa, Florida; the Baytown Bridge in Baytown, Texas; the Talmadge Memorial Bridge in Savannah, Georgia; and, for the past three years, the New Baldwin Bridge in Connecticut. He is Project Manager for Parsons Brinckerhoff Construction Services, Inc. He became a U.S. citizen in 1992 and resides in Essex, Connecticut, with his wife, Jacqui, and two children.

About the Illustrator

EDITH TWINING is an author/illustrator who spent the summers of her childhood in Old Lyme where her family has lived for three generations. She started drawing scenes of the Balwdin Bridge as soon as work began and is considering writing and illustrating a children's book about bridge construction. Edie wrote and illustrated *Sandman*, a children's book published by Doubleday in 1991. A graduate of the Boston Museum School, she lives and works in Boston but is frequently in Lyme visiting her parents with her husband and two children.